WOMEN AND WORK
IN SOUTH ASIA

NF

WOMEN AND WORK IN SOUTH ASIA

Regional Patterns and Perspectives

Edited by
Saraswati Raju and Deipica Bagchi

London and New York

First published 1993
by Routledge
11 New Fetter Lane, London EC4P 4EE

Simultaneously published in the USA and Canada
by Routledge
29 West 35th Street, New York, NY 10001

© Saraswati Raju and Deipica Bagchi

Typeset in Garamond 10/12 by Florencetype Ltd, Kewstoke, Avon
Printed and bound in Great Britain by Biddles Ltd, Guildford and King's Lynn

British Library Cataloguing in Publication Data
A catalogue record for this book is available from the British Library

Library of Congress Cataloging in Publication Data has been applied for
JK
ISBN 0–415–04249–6

Limits of my language are the limits of my world. And it means the most difficult thing of all: listening and watching in art and literature, in the social sciences, in all descriptions we are given of the world for the silence, the absences, the nameless, the unspoken, the uncoded – for there we will find the true knowledge of women.

Adrienne Rich in Lies, Secrets and Silences, p. 245

CONTENTS

CONTENTS

**Part III Gender ideology, power and powerlessness:
empirical observations**

Part IV Epilogue

FIGURES AND TABLES

FIGURES

TABLES

CONTRIBUTORS

Meena Acharya received her Ph.D. from the University of Wisconsin, USA. She has had a long career with the Nepal Rastra Bank and worked with the World Bank's Development Economics Department for two years. Her research interests are the rural labour market, women's participation in the labour force and household decision-making. Her works have been primarily in reference to Nepal. She has been actively associated with the World Bank's project on women in Nepal. Currently, she is the Executive Director of a prestigious private research trust, the Institute for Integrated Development Studies, Kathmandu, Nepal.

Huma Ahmed-Ghosh received her Ph.D. in anthropology from Syracuse University, USA. Her dissertation was on the impact of agricultural development on women's status and work patterns in Palitpur, India. She has published and presented several papers on this theme.

Joyce Aschenbrenner received her Ph.D. in anthropology from the University of Minnesota, USA. Her dissertation on endogamy and social status in a west Punjabi village was based on a year's fieldwork under a Fulbright Fellowship. She has done research on black families in the USA and is also involved in an oral family history project in Southern Illinois. She completed a curriculum project entitled 'Equity Around the World', focusing on South Asian women for the St Louis School District. She has directed the Women's Studies Programme at Southern Illinois University at Edwardsville and was Chair of the Anthropology Department. She is presently Professor Emerita.

Deipica Bagchi Ph.D. is an Associate Professor of Geography at Southern Illinois University, Edwardsville (SIUE), USA. Her research of the past fifteen years on rural women's role in agricultural modernisation, rural energy and environment conservation has led to the publication of several articles in national and international journals, and participation in three of the four International Interdisciplinary Congresses on women. She is an active member of the International Geographical Union (IGU)'s Commission on Changing Rural Systems and the Association of American Geographers, specialty Group on Geographic Perspectives on Women and Rural Development. She teaches a course on 'Gender and geography'. Currently, she is the director of the Women Studies Programme at SIUE.

Kalpana Bardhan Ph.D. was a Lecturer in the Economics Department in the University of California, Berkeley from 1977 to 1987. During 1982–7, she also held the position of research consultant for the ILO, the ILO-ARTEP and for the United Nations University (Tokyo). Her research at ILO resulted in several working papers on women's work, rural labour markets and poverty-related employment problems. She has also written and published extensively in journals of national and international repute on these themes as well as price relationships and responses in Indian agriculture. Currently, she is engaged in full-time research.

Sabine Felmy is a Master of Arts in English, philosophy and educational Science. She has conducted extensive field trips in South Asia including numerous visits to the northern areas of Pakistan. Within the last ten years, she spent almost two years in Hunza and published a book on folk tales from Hunza in 1986.

Doranne Jacobson Ph.D. has conducted extensive anthropological field research on women and their seclusion in a village of Madhya Pradesh, India, over the past three decades. Her research on the roles of women in South Asia has resulted in the publication of a book and more than two dozen articles in books and national and international journals. Many of her photographs have appeared in exhibits and publications around the world. Her most recent volume, *India*, published by W.H. Smith, is lavishly illustrated with her photographs. She is Director of International Images, Springfield, Illinois, USA.

Joan P. Mencher Ph.D. is a Professor of Cultural Anthropology at Lehman College and City University of New York, USA. She has been a consultant with UNFPA, India (1990), and World Bank, Sri Lanka (1989–90). Her research for the past fifteen years has been in the field of women in agriculture especially in rice cultivation and women's life-history profiles, and social change in South India. Based on her fieldwork in these areas, she has published extensively in national and international journals of repute. Her books include *Social Anthropology of Peasantry* (ed.) and *Agriculture and Social Structure in Tamilnadu: Past Origins, Present Transformation, and Future Prospects*. Currently, she is a Co-ordinator of Women's Studies at Lehman College.

Saraswati Raju Ph.D. is an Associate Professor of Social Geography at Jawaharlal Nehru University, New Delhi, India. Her interest in gender issues began when she was a post-doctoral fellow on an Indian Government Fellowship at the Department of Geography, Syracuse University, USA during 1978–81. Since then her research and publications in Indian and international journals include analyses of Indian data on female literacy and labour participation. She has participated extensively in national/international seminars and workshops on this theme. She has recently introduced an M.Phil. course on 'Regional dimensions of female labour force with special reference to India', the first of its kind to be taught in geography in India. She is the representative for the South Asian region on the International Geographical Union (IGU)'s Commission on Gender and Geography.

Joke Schrijvers Ph.D. is Professor of Development Studies at the Institute of Development Research Amsterdam (InDRA), The Netherlands. In 197,6, she helped in establishing the Research and Documentation Centre on Women and Autonomy at the University of Leiden, followed by anthropological action-research in Sri Lanka in 1977–8 on the effects of planned and unplanned changes in the lives of peasant women. She supervised research projects in Sri Lanka, Indonesia, Egypt and Mexico. She has published works on Sri Lanka and feminist anthropology, and her current research interests include displacement and violence (*The Violence of Development*, 1993).

PREFACE

This book is an outcome of a growing concern about women's work that often remains invisible in official statistics and development research, more so in the South Asian context. This is partly because of the inadequacy of our national data systems which fail to capture women's work and partly because existing sociocultural constraints restrict women's participation in economic activities outside the familial domain.

During and before the gestation of this book, both editors carried out fieldwork and research, published papers, introduced courses and supervised students' research endeavours in their respective universities on the theme of women and work. This long exposure made them realise that the observed pattern of female labour participation in South Asia has distinct spatial dimensions, many of which cannot be understood and explained in terms of economic rationale alone and that the region-specific context defining women's role remains vitally important.

All too often, this context is defined at an aggregate level without any reference to sub-regionally and locally induced religious, cultural and societal constraints on gender relations and therefore women's work. Individual case studies, on the other hand, fail to situate themselves in a broader geographical framework. This book is an attempt to fill this gap in the literature by integrating the different scales of analysis (and thus different methodologies) to make the volume not only a contextually specific collection, but also one where the contextual specificity has been placed in a wider perspective.

The overall scenario for South Asia is presented with the help of official data. This is inevitable because of the immense canvas involved. However, a few chapters are based on micro-scale primary data. These clearly bring out the 'public' and 'private' domains of women's work and the glaring shortcomings of official data-collecting agencies provide counter references which make it possible to examine various propositions back and forth at both levels. The editors feel that this approach greatly enriches this collection.

The reader will note the geographical spread of authors with varying academic orientations working in the same general field of feminist research

xvii

lending a certain thematic coherence to the volume: those with non-South Asian backgrounds, wary of the potential pitfalls of an 'outsider' researching in alien cultures, bring in, at the same time, the ever elusive 'objectivity' difficult if not impossible for a native, while the insights and subtle nuances captured are claimed by the indigenous authors.

Admittedly, there are a few omissions. The rural component of the female labour force has been covered more extensively as compared to urban counterparts. Pakistan should have been covered in the macro section. We feel a little deprived in that we were not able to tap any latest writing on Bangladesh. The role of state and state policies have an important bearing upon women's work which a political scientist could have brought into a sharp focus. However, we have not been able to include any study focusing on this aspect. We regret these omissions and hope to include what we missed this time in a revised edition, if there is one.

We thank the Centre for the Study of Regional Development, School of Social Sciences, Jawaharlal Nehru University, New Delhi, India and the Department of Geography, School of Social Sciences, Southern Illinois University, USA for their help. We acknowledge the help we received from the anonymous reviewer's extensive comments on the earlier draft in making our book tighter and better targeted. We are grateful to our publisher, Tristan Palmer, for his encouragement and support, and to our desk editor, Diane Stafford, for her overall editorial assistance. Eleanor Rivers remains in our memory as the first ever link in our association with Routledge. Our heartiest thanks, however, go to the contributors whose patience and understanding held firm across space and time to make this volume a reality.

INTRODUCTION

Saraswati Raju

ON WORK

The population that has the ability and a willingness to work constitutes the labour force. The age composition of a given population would largely determine the proportion of those who would be in the labour force. However, who actually gets included in the workforce would very much depend upon how 'work' is defined in such accounting. This is not to deny some basically varying structural differences between the countries affecting actual work participation rates, but much of the variation therein results because of the conceptual and measurement problems implicit in the evaluation of work within the labour force, particularly with regard to females. Why this should be so is an intriguing question.

Broadly, there appears to be a general gradient of decreasing female employment from industrial to agricultural countries. These generalised patterns have within themselves much variation, being most marked among primarily agrarian countries. Cutting across these economic groupings are broad regional-cultural tendencies, notably in lower rates of female participation relative to males. South Asia, the subject of this study, clearly falls towards the lower end of the scale of female workforce participation though there exists considerable variation within the region itself. Thus, the northern region resembles the West Asian and North African Arab countries in having low participation rates, whereas central and southern parts of South Asia have high participation rates like Southeast Asian countries.

In itself low, the average rate for female workforce participation noted in India, i.e. 23 per cent, is significantly higher than that of Bangladesh and Pakistan (7 and 13 per cent respectively). Nepal and Sri Lanka (43 and 29 per cent), on the other hand, are at the top of the scale (United Nations 1991; also, Kazi 1989: 35–41).[1]

Because of the uneven way work gets defined, labour data on one country from various sources vary tremendously. Thus, estimates derived from the Pakistan Fertility Survey, the National Impact Survey and the Pakistan Contraceptive Prevalence Survey record rates of female work participation

1

which are nearly double the rates derived from the country's official statistics (Irfan 1981; Shah 1986; Afzal 1989; Kazi 1989: 36). In India, the National Sample Survey Organisation (NSSO)'s estimates are higher as compared to the census data (Nayyar 1987). In the case of Nepal, a more liberal definition of gainful participation yields a higher share of work for women as compared to men (Acharya and Bennett 1982). For Bangladesh and Pakistan, rates of work participation by rural women are not only very low but they are also lower than the corresponding urban rates, although slightly so in the case of Pakistan (only in 1981) in contrast to the general pattern observed in South Asia (Kazi 1989: 36; Shah 1989: 157). Such discrepancies between data sources (and over time) reflect the inadequacies of female labour force statistics in South Asia.

Traditionally, work has been defined as activities for 'pay or profit'. In official statistics also, the term 'economic activity' rather than 'work' is used to maintain the consistency with the concept of economic production (in the UN System of National Accounts). Although the definition of economic activity has been broadened over the years, the underlying emphasis on paid work has not changed. This essentially income-oriented approach to the definition of work can capture only those who are in a) wage and salaried employment, b) self-employment outside the household for profit, and c) self-employment in cultivation and household industries for profit. This definition is inadequate in a partially commoditised economy where a significant proportion of goods and services are produced for self-consumption (Duvvury and Isaac 1989: 1). Exclusion of such activities from the definition of 'work' affects male workers also, but prevailing sex segregation in work results in a much higher undercounting of females (Boserup 1970; Boulding 1976: Dixon 1982a; Banerjee 1985; Singh and Kelles-Viitanen 1987; Beneria 1988; Krishnaraj 1990). To counteract this the Indian Census counts cultivation of crops as 'work' even if it is for self-consumption. Despite this the gulf between the male and the female participation rates exists because even in agriculture much of the related activities resulting in produce for self-consumption, primarily undertaken by women, are invisible and cannot be captured by the existing concept of work. However, as the following discussion shows, conceptual clarity alone does not automatically ensure an accurate estimate of the female labour force. This situation is not unique to South Asia as these problems plague the labour force statistics everywhere in the world (Dixon 1982a: 539; United Nations 1991).

As pointed out by Beneria (1988), in traditional economics, activity is evaluated in terms of capital growth and accumulation which invariably places quantitative relations in commodity production as the supreme factor. Since the market is a place where these relations are articulated, it becomes the formal expression of economic activity. The result is a logical exclusion of all activities outside the market mainstream as 'peripheral' and 'non-economic' (Beneria 1988). Since most of the pre-harvest and post-harvest operations in which women indulge are carried out at home, a large number of self-employed women are excluded from the count.[2] In addition, allied agricultural activities such as dairying, poultry farming, along with survival tasks of firewood/fodder collection and procurement of water are not considered work if they are for self-consumption. As compared to the census, in India, NSSO has been more sensitive in its attempts to capture women's contribution to work by adopting multiple approaches to define economic activities. For example, its 32nd, 38th and 43rd rounds (conducted in 1977-8, 1983, 1987-8) introduced innovative measures such as inclusion and division of domestic workers into two categories: first, those engaged in 'domestic duties' only (activity code no. 92) and second, those who 'attended domestic duties and [were] also engaged in free collection of goods (vegetables, roots, firewood, cattlefeed etc.), sewing, tailoring, weaving etc. for household use' (activity code no. 93) (NSSO 1988). For this an exclusive set of probing questions for those who were categorised as 'usually engaged in household duties' was adopted to elicit information on their participation in certain specific activities for household consumption.[3] By including the latter category the female labour participation rate not only comes closer to that for men, but the coefficient of variation, i.e. the variation in female labour force participation across the Indian states also becomes considerably smaller (Sen and Sen 1985). This establishes that, all over the country, for females, expanded household activities are quite common. However, persons returning code no. 93 as their main activity have been treated as 'nonworkers' in both the 32nd and 38th rounds as against the earlier 27th round where women in this category have been treated as 'workers' (Krishnamurty 1990).

Ironically, activities under this category may be so overwhelmingly burdensome that women may not be able to spare time to engage in activities which will place them in the formal category of workers. In fact, an analysis of the 'reasons for attachment to domestic duties' as revealed by these rounds shows that for more than 90 per cent of women this indeed is the case while non-availability of work was responsible for only 3 per cent of the women remaining outside the formal labour force (Kundu and Premi 1992).

With the exception of agriculture, the census and even NSSO in India do not consider a) primary activities (other than agriculture) i.e., animal husbandry, mining/quarrying, fruit and fuelwood gathering for household consumption, b) processing of primary product for household consumption, and c) own account production of fixed assets as economic activities although a significant number of women engaged in such activities have been producing what the UN System of National Accounts term as 'economic goods and services'.[4]

Bringing in the unpaid work adds yet another dimension to female labour, i.e., longer hours of work for them. In Nepal, males and females spend almost identical hours in economic activities, but females have eight additional hours a week as compared to their male counterparts when non-market economic activities are considered (United Nations 1991).

Even if the 'household consumption' is accepted as cut-off point between work and non-work there are many activities in agricultural households which are carried out at home but which are not for household consumption. A housewife preparing meals for the consumption of agricultural labourers in the field (forming part of the wage deal) or snacks for sale by petty hawkers on streets in urban households are cases in point (Pariwala 1985; Saradamoni 1987: WS 4; NIUA 1987; Papanek 1989).[5] Similarly, a potter's wife who collects clay and processes it before it goes to potter's wheel and afterward helps in lighting a kiln for baking the pots is most likely to be recorded as a non-worker! The same tendency is observed in handloom production, horticulture and sericulture (National Commission 1988). A study from Bangladesh reports that participation of women in pisciculture is 'substantial' as about 30 per cent of women in rural coastal areas are directly or indirectly engaged in small-scale activities related to fisheries (Raja 1988; Sharma and Kumar 1988; National Commission 1988). Huma Ahmed-Ghosh's study in this volume refers to a north Indian village where women from higher castes have to undertake jobs including tending livestock and cleaning of cattlesheds which were previously done by paid labour. Such conversion of paid jobs into unpaid ones, essentially taken over by women, is quite common in villages of India where lower castes are experiencing upward mobility.[6] These activities are clearly outside the purview of so-called 'household consumption'. However, the proximity and integration of these activities with domestic work makes them highly invisible (Bardhan 1985; Shah 1986; Beneria 1988; Duvvury and Isaac 1989).[7] Dixon has commented on the demarcation line between domestic production for the household's own consumption and sale or exchange oriented economic activity by stating: 'where the household ends and the farm begins, in spatial terms, or where housework ends and production begins, in economic terms' is necessarily an arbitrary decision (Dixon 1982a: 542).[8]

The suppression of accounting of labour force participation by national and international agencies is because they count the stock of workers at one point of time rather than through the days and seasons, that is, the flow of productive labour (Dixon 1982a). This lacuna emerges more glaringly through time allocation surveys. However, gender-aware studies on the pattern of time allocation in rural households of South Asia are virtually non-existent except for a few sample surveys carried out in different parts of the region such as Karnataka, Rajasthan and West Bengal in India (ASTRA 1981; Jain 1985), and a few on Bangladesh.[9]

One of the ways to deal with the household and non-household dichotomy is to count all household work such as cooking, cleaning, tailoring, child and elderly care as economic activities. The underlying argument would be that these activities have a price tag, i.e., they can be substituted for goods and services which are otherwise paid for. On the flip side, however, this would mean inclusion of all except children, sick and elderly as workers with no temporal or spatial variation (Dixon 1982a: 542). While this certainly makes the crucial role of women in the economy visible, it also clouds the real issues and a certain complacency may ensue because female labour would then no longer be invisible. However, apart from cooking, cleaning, childcare, and looking after the aged and the sick which may fall within the 'pure' domestic sphere, Maithreyi Krishnaraj (1990: 2664–72) has identified several categories of work predominantly performed by women in rural and urban India which clearly have components that are not 'pure' domestic work. They are: a) self-employment in cultivation for own consumption, b) subsistence dairying and livestock rearing, fishing, hunting, and cultivation of fruit and vegetable gardens, c) fetching fuel, fodder and water, repair of dwellings, making of cow-dung cakes, and food preservation. In addition, women's work in 'informal' healthcare is particularly invisible in official statistics (and social research). Moreover, many women are able to participate in the formal labour market because other women (usually kin) are looking after dependent children.

Reference period, fluctuations and problems of comparability

Another source of under-enumeration is the reference period for recording work status: one has to work for a stipulated period in order to be classified as workers. This affects female workers, especially in rural areas because their work is largely seasonal (Deere and Leal 1982: 12–13). In the Indian situation, for example, NSSO used to have a reference period of one week preceding the enumeration day during which the person should have worked for at least one hour on any day to be placed as a worker under 'current weekly status'.[10] This, however, precludes many women if the survey is carried out during a slack period.[11] From 1972 to 1973, the organisation started using the concept of 'usual status', i.e., within a reference period of one year preceding the date of enumeration. The person is expected to be engaged in some activity for a long period of time in the past and is likely to continue in future to be classified as worker in this category. In addition to these two categories, there is the 'current daily status' to capture the activity pattern of that segment of population which is in the unorganised sector whereby a person may be pursuing more than one activity during a week or sometimes even during a day. Some of these activities may not be officially defined as 'gainful'. As such this status is assigned to those persons who have worked for four hours (half day each up to a maximum of two types of activities) or more on each day during the reference week. The unit of classification is thus 'half day'. Understandably, the female work participation drops sharply from usual status to daily status, more so for rural than urban females (Bashir 1988: 9).

In India, the main source of data on the female labour force is the census which is carried out every ten years. The definition of work has been the same in the three census years starting from 1961 (all the previous censuses had concept of 'income' or 'economic independence' as the basis for identification of work). That is, all those who are physically or mentally involved in economically gainful activities have been defined as workers. The problems of comparability are induced because of changes in the reference period particularly in the case of female workers (Anker et al. 1987; Reddy 1988). In the 1961 and 1971 censuses, the dual approach of usual status and current status was adopted. That is, for the concept of 'usual status' the reference period was one year for seasonal work in l961 and 1971 whereas for regular work, it was one fortnight in 1961 and then reduced to one week in 1971. In addition to the shorter reference period for regular work in 1971, the question on work status was 'what is your main activity?' (more often eliciting 'housewives' as an immediate response). Given the nature of female work, the use of main activity as the starting point, as against the 1961 census when respondents were first asked to identify their work and later divided into main and marginal workers, led to exclusion of most of the female respondents and they were treated as non-workers in the 1971 census. This resulted in a sharp, albeit spurious, decline in the female participation rate from 27.93 per cent in 1961 to 14.22 per cent in 1971.[12] In the 1981 census the usual status approach was adopted uniformly for all work. The question asked was, have you 'worked any time at all last year?' In cases of affirmative answer, the respondents were asked what their main activities were. Those who worked for at least 183 days (six months) in the preceding one year were treated as main workers and those who did not work for this period were treated as marginal workers. This approach was adopted to make the 1981 data on main workers comparable with 1971 data. Main plus marginal workers in 1981 could correspond to the data on workers in 1961. However, scepticism about such a comparability exists in literature (see, Duvvury 1989a; Unni 1989). The use of key words and wording of questions thus assumes importance (United Nations 1991). The 1991 census adopted a similar approach, but extra care was taken to tap the female workers (Office of the Registrar General and Census Commissioner 1991b and c).

In South Asia, the other country to adopt this dual classification is Sri Lanka. It may well be that their relatively high levels of female participation in the labour force is a product of this practice as their intermittently employed marginal workers are mostly female. It is sometimes argued that the abysmally low level of labour participation by women in Bangladesh and Pakistan might be attributed to 'the absence of a long reference period' in their surveys or census definition (Jose 1989: 5). However, the participation rates among women in Pakistan as reported by the Census-cum-Housing,

Household Economic and Demographic Surveys (HED) of 1972/3 which has a reference period of only one week are higher (5.1 per cent) than those reported in the following census year in 1981 (3.2 per cent) where there was no specific reference period at all (for data, see Shah 1986: 268, 1989: 155). Since data from other countries on the basis of extended reference period show a reverse trend, the relatively lower participation rate in Pakistan for 1981 may or may not be a part of that unresolved debate on generally declining female participation in the labour force over time. Inadequate information at this juncture makes any definite statement redundant. It may be noted, however, that in the case of married women there, the extension of reference period from one week (Labour Force Survey) to one year (Migration Survey) drastically reduced the share of unpaid family helpers and to some extent that of self-employed persons. This may be because a longer reference period for recording some activity may have created an impression of having a steady job in the minds of respondents (Irfan 1981: 5).

Nepal, on the other hand, includes, in the economically active population, only those who are ten years of age and above and have been working (for pay or profit) for at least eight months either continuously or sporadically. The unpaid family workers are also included if the time thus spent is at least one-third of normal working time. The inclusion of the ten years and above age group raises the participation rates. However, the emphasis remains on economically gainful paid work, although in 1991 the manual for enumerators contained special instructions to extract information on work especially from females by asking probing questions (National Planning Commission 2047 (Nepali Calendar Year for 1991)).

Going back to the Indian situation, the 1981 census, for the first time, put explicit emphasis on 'probing questions, particularly in the case of women, to find out if they have any economic activity, even if it is marginal, apart from household duties' (Office of the Registrar General and Census Commissioner 1981: 15). However, these instructions were not built into the questionnaire and therefore its effect on actual recording of data is difficult to comprehend (Duvvury and Isaac 1989).

Although the concern in the feminist literature has not really gone beyond identifying and highlighting the whole range of tasks done by women and the need to include that as 'productive work', there is little doubt that the 1980s witnessed an express interest in women's issues and impassioned lobbying by scholars, women's organisations, activists and others with political clout. One of the fallouts has been a growing sensitivity towards female contribution in non-wage activities. For example, the Indian census in 1991, has attached an explicit rider in the identification of workers to include 'unpaid work on farm or in family enterprise' (Office of the Registrar General and Census Commissioner 1991b; also Krishnaraj 1990). This is expected to capture women's unpaid family labour.[13]

The NSSO data are not free from problems of temporal comparability either. In various quinquennial employment surveys conducted by this organisation, the activity status 93, i.e. persons attending to 'domestic duties and also engaged in other primary activities which are carried out as a part of their household chores', has been treated as 'worker' in earlier rounds, whereas later these activities were excluded from work. That even after making adjustments for these changes the estimates fail to capture the 'actual female participation rates' is a separate issue (Kundu and Premi 1992).

The perceptional, enumerators' and respondents' biases

The conceptual and definitional inadequacies as outlined above are simply articulation of deeply embedded societal biases regarding the role of women who are primarily seen as wives and mothers while the man is the breadwinner. A logical outcome of this ideology is that man is linked with the public domain whereas woman's place is inside the home, the proverbial familial domain. However, this is not a phenomenon unique to South Asia. In a shift from preindustrial to industrial stage, as production began to move out of households into factories, offices and stores, workers who got paid were clearly identified as workers. Their's was socialised labour which attained greater value as opposed to domestic labour with no exchange recognition. The separation of work space from home elbowed women out of the public domain as their spatial mobility was constrained by their reproductive and household responsibilities.[14] First, the home-based activities were relegated as secondary; second, their restricted mobility seems to have shaped the societal norms about the role of women. So pervasive are these perceptions that women themselves undermine their work status (Agarwal 1985; Anker *et al.* 1987; Devadas *et al.* 1988; Patel 1989: 15–16).[15] Further, in the sex-segregated society of South Asia, more often the male investigators are not even allowed to talk to women respondents and reporting by males is full of biases. In two national household surveys of 1968–9 and 1975 in Pakistan the activity rates when reported and recorded by women were found to be twice that of the 1961 census rates (Shah and Shah 1980; Irfan 1981; Kazi 1989). Anker has shown how the sex of the enumerator and of the respondent makes a significant difference to the outcome of such inquires (Anker 1983). In a later study conducted in Uttar Pradesh (India), however, the survey results were found not to be significantly affected by the sex status of the enumerator or respondent (Anker *et al.* 1987). This could be because of better trained male enumerators and a small-size sample. Nevertheless, even these may not successfully control secondhand and less than candid responses by women in the presence of male members. Even in the absence of males, female responses are often moulded by what they know their menfolk want them to say. This situation may, however, differ from one familial context to another (Asuri and Mahadevappa 1988: 4–5). As such, generalisations based on small-scale samples cannot be applied to large-scale operations or other locations (Krishnamurthy 1985; Duvvury and Isaac 1989).

Due to societal biases whereby females are recognised only as supplementary earners, even the incidences of female-headed households go unrecognised and undercounted while the phenomenon is on the rise in the subcontinent (Jain and Chand 1982; Krishnaraj and Ranadive 1982; Visaria and Visaria 1985; Saradamoni 1988). In 1982, more than 15 per cent of total households in Bangladesh were reported as headed by women (Government of Bangladesh, Bureau of Statistics 1986). In India, 10 per cent of households were recorded as headed by women in 1961 and 1971. In 1981, this percentage was reported to be 16 (Visaria and Visaria 1985). The notion that women are only supplementary earners is flawed on two accounts: a) even in a family where both man and woman are earners, the responsibility of maintaining the family may in fact lie with the woman, and b) among the poorer households, the distinction between 'below or above the poverty line' may hinge precariously on women's income (Gulati 1976, 1978, 1984; Chakravarty and Tiwari 1977; Kumar 1978; Bhatty 1981, 1987; Acharya 1983; Dasgupta and Maiti 1986; Saradamoni 1987; Mencher 1988).[16] In fact, in the sociocultural context of South Asia, it is extremely difficult to distinguish between the head of the household and family, and more often the eldest or the only male family member is identified as 'the household head' by the respondent even if no monetary contribution by the person is involved. In several cases, women may even work to sustain their families while the male members are either gaining experience, education or waiting for the right job to come along (Karlekar 1979; Kasturi n.d.). In many cases, the concept of 'household head' is used more as a social construct rather than involving monetary connotations.

In sum, two focal points emerge. One is the concern over biased and incomplete reporting of female workers which can be rectified even within the existing definition of work. The other relates to careful examination and recognition of those aspects of 'non-work' which need to be included as 'work' keeping in view the complex socioeconomic realities of women's lives in different cultural contexts.[17] In order to deal with the first problem, it is necessary to counteract the ideological undervaluation of female labour through a gender awareness campaign to sensitise not only the respondents (including women), but also those who are involved with the recording of such data to a wide range of activities in which women are engaged. The 1991 census in India made some progress in this direction when at the insistence of UNIFEM some changes in the training manual for enumerators were agreed upon. A pictorial description of various economic activities undertaken by women has also been prepared to be displayed at every numeration block. The second proposition, however, calls for expanded definition of work. From this perspective, applying international recommendations at individual nation's level has been rather difficult apparently because of a) efforts and operational costs involved in such an exercise, and b) the problems of temporal comparability. In a way, the unwillingness on the basis of cost factor (in terms of both time and money) is linked with the obscurity and low value generally attached to female labour in predominantly patriarchal structures. If to begin with, the female labour is perceived to be subsidiary and subordinate to males' in nature, the logical consequence would be not to acknowledge it on some pretext or other. However, more often the reluctance is due to apprehension on the part of data collecting agencies of rendering the data incomparable over time. This is a rather shaky argument because even without the expanded definition, the ad-hocism adopted in defining the labour force in different censuses as well as reshuffling certain activities from one category to another and their inclusion/ exclusion from one count to the other has made temporal comparisons possible only after tedious adjustments, especially in the case of female labour. If the concern is not to compound the complications further, instead of the traditional division between workers and non-workers, data can be collected by adopting multiple definitions of labour force which would cover paid labour (as has been traditionally defined in a given country so as not to lose comparability with earlier data) as well as the labour force captured through an additional questionnaire (according to an expanded and more realistic definition, the one which can be in keeping with the international recommendations as well) that excludes only purely reproductive and social activities (Anker 1983; Nayyar 1987; Dixon-Mueller and Anker 1988; Duvvury and Isaac 1989; Kundu and Premi 1992). The results can be made simultaneously available. If the logistics of such an exercise do not make thorough coverage possible, data can be initially collected on sample basis. Some of the data on labour force is qualitatively better than the population census, as is the case with the NSSO data for India, and the concerned authorities should be urged to provide the data at disaggregate (district) level as suggested by the Working Group on District Level Planning set up by the Planning Commission of India as against the current practice of publishing them at state level only.

Whether child-bearing and child care, cooking and cleaning, currently not considered at all in either national accounts or labour force statistics, should be counted as 'work' or not is that part of the debate which remains largely unresolved. But as mentioned earlier, an affirmative answer to this question has its own catch.

If all the ambiguities in the data system are somehow removed, would it mean that the variation in the levels of female labour participation across space would disappear? Interestingly enough, the inclusion of domestic tasks such as collection of fuel, fodder and water does reduce the coefficient of variation in female participation rate across the Indian states from 0.452 to 0.124 (Sen and Sen 1985: WS 52). This is not surprising because these tasks everywhere have almost always been seen as associated with female domain

(Beneria and Sen 1988). It may be argued, therefore, that some apparent regional patterns are simply artifacts of the way in which female labour is defined.

In terms of conventional female labour force, however, conceptual ambiguity and restricted coverage is only a part explanation for observed variations in female labour force participation. Much more difficult to comprehend are the ways in which the labour market becomes gendered.

CONTRADICTIONS AND RESOLUTIONS

The inherent and exogenous difficulties with which the entire question of female labour participation, its identification and presentation is confronted are further compounded by societal norms and biases, which indirectly impinge upon prevailing economic mores, and accord women their position in the caste and class hierarchies. This results in distinct regional and sub-regional variations within and between the countries in South Asia (Agarwal 1985: A159; Bardhan 1985: 2262; Saradamoni 1987: passim; Acharya present volume). This section attempts to capture some of these complexities. In doing so, although we have tried to separate the constituent factors under different sub-sections, a certain degree of overlap could not be avoided because of the intertwined nature of some of the variables. Further, the broad generalisations relate more to what may be termed as the 'supply side' of the equation. The factors affecting the demand side, it may be pointed out, have been touched upon only to the extent that they help us understand the ongoing processes in the labour market.[18]

Agroecological context and labour

Since countries with lower levels of female work participation also have agriculture as the leading occupational category, the variable pattern of labour force participation is attributed to geographically distinct regions where supply and demand for female labour is generated by ecological differences influencing farming intensities and cropping patterns (Miller 1981; Rosenzweig and Schultz 1982; Chen 1989). Without going too deeply into the details of different types of agricultural ecologies and their requirements for female labour, a few propositions can be explored. There exist some basic differences between labour demands of dry-field plough agriculture and intensive wet cultivation. Within the realm of plough agriculture, areas of wet-rice cultivation are expected to show a higher participation of females in the labour force because rice cultivation makes heavy demands on labour in transplanting and weeding operations (Boserup 1970). On the basis of this essentially Boserupian model, several scholars have hypothesised a rice-wheat dichotomy in agriculture as explaining the

11

variation in female work participation across regions (Bardhan 1974; Bardhan 1984; Mencher 1977).

Such an ecologically oriented explanation for the exclusion/inclusion of women from and in agriculture is, however, open to question. The argument regarding the effects of different 'demands' and 'requirements' for female labour under different crops, and technically determined correlation between plough cultivation and the actual level of female employment as envisaged by Boserup, though an appealing one, needs further examination.[19] The relationship, it can be pointed out, is obscured in the contemporary South Asian context. The overall female workforce rates are relatively higher in the south of India, which is essentially a rice-producing region, but the western and central states despite their relatively higher workforce participation of females are not producing rice.

On the other hand, women are segregated in some wet-rice areas, such as Bihar and West Bengal. Miller mentions the paradoxical case of West Bengal where the task of transplanting, which has traditionally been associated with females, is done by males. Similar deviations from standard practices in agricultural operations are also noted elsewhere (Mencher present volume; Horowitz and Kishwar 1984: 10; Harriss and Watson 1987: 95–7; Saradamoni 1987: WS 3).[20] Moreover, it is irrigated paddy which is characterised by higher incidence of female workers as compared to paddy areas *per se* (Ryan and Ghodake 1984; Sen 1987).[21]

In Orissa, a predominantly rice-growing state in eastern India, the female work participation rate is lower than the neighbouring southern states, being more in accordance with the wheat-growing north (Agarwal 1984, 1986b; also, Jain 1985).[22]

Several scholars have argued that a higher percentage of area under rice depends upon assured irrigation facilities and leads to higher income which is likely to result in low female work participation. Accordingly, to rice and wheat must be added another agroecological zone of coarse-grain cultivation. The general argument is that the income from agricultural activities in dry cultivation is so inadequate that females may have to work to supplement the family income (Reddy 1975; Sen 1982, 1987; Chatterji 1984; Chen 1989; Duvvury 1989a; Saxena 1990).[23] In Andhra Pradesh (India), however, the work participation rates tend to be lower for the poor in dry areas while in irrigated areas they increase with increase in per capita income (measured in terms of expenditure) (Reddy 1991: 42).[24]

In yet another set of observations commercial cropping pattern has been posited as assuming importance as far as absorption of female labour is concerned. In India, the introduction of tobacco cultivation in the early 1920s has been found to have resulted in a tremendous increase in female labour as a range of operations from transplanting to curing was performed by females (Mies 1984). Similar observations have been made from cotton-growing areas in Maharashtra, India and tea plantation areas of Sri Lanka

(Reddy 1981; Acharya and Panwalkar 1988). However, as pointed out by Duvvury, the relative impact of cultivation of cash crops depends on the circumstances under which it is started. In areas of subsistence farming a labour-intensive cash crop may lead to a significant increase in female labour. However, if the existing crops are already labour intensive such as paddy then the increase in demand for additional labour would depend upon the relative labour intensities of the respective crops (Duvvury 1989b).

Without delving further into the debate it may be logical to conclude that the cropping pattern is just one specific factor in the whole nexus of socioeconomic considerations affecting female labour and any generalisation based on this in isolation is bound to be misleading (Nagaraj 1989: 110).

Gender, poverty and labour

In other contexts also strong association between poverty (usually the incidence of landlessness taken as surrogate variable to denote poverty) and rural female work participation has been posited (Chatterjee 1984; Jain 1985; Sen 1985; Agarwal 1986b; Nayyar 1989). That rural impoverishment should indeed drive the potential workers to agricultural wage work seems quite rational. However, in the South Asian context unidirectional association between female employment and poverty is not readily open to statistical analysis (Duvvury 1989b: 90). Boserup refers to the state of Uttar Pradesh in northern India where the social resistance to female employment in the public domain is so rigid that even agricultural labourers who are usually the 'poorest of the poor' keep their women at home (1970: 75).

The NSSO data on labour force participation grouped according to the standard of living (calculated on the basis of per capita expenditure for the year 1977–8; see, NSSO *Sarvekshana* 1986: Table 1) show differentiated patterns between Bihar, Haryana, Punjab, Uttar Pradesh and West Bengal in northern India, and Andhra Pradesh, Tamilnadu, Gujarat, Madhya Pradesh, Maharashtra and Rajasthan in southern and central India. In the former group, not only are the overall female labour force participation rates lower than the average for the country, but they also vary tremendously between the rich and the poor females. Since male labour participation rates do not vary that drastically, the gender gap in labour widens significantly with increase in income. In the latter states, the overall female labour force participation rates are above the national average and the extent of differentiation across various expenditure classes is much less pronounced resulting in a much more gradual increase in the gender gap. Intriguingly, despite relatively higher labour participation rates for the poor females in the north, in some cases their participation rates fall short of that of females in the highest per capita expenditure group in southern and central Indian states (Nagaraj 1989: 115–16 particularly Table 4.4). A more recent World Bank report on *Gender and Poverty in India* observes that the most prosperous northern states like Haryana and Punjab share low female participation in labour force with the eastern states of Bihar and West Bengal where the incidence of poverty is extremely high. Among major states, Maharashtra, the third ranking state in terms of per capita income, is characterised by the highest female labour work participation rate (World Bank 1991: 27–8; Raju 1992).[25] Even in 1991 (census data), there is no consistent link between per capita income and female work participation rate. However, analyses at state-level aggregation have to be seen more as suggestive than conclusive and there exist a few micro-level studies to indicate some association between income and female labour participation rates (Parthasarathy and Ramarao 1974; Sawant and Dewan 1979; Jain 1985).

Withholding the overall low female participation in the north Indian Plain, in terms of sectoral break-up and the temporal change, between 1971 and 1981 the increase in the number of female agricultural labourers per 100 male agricultural labourers (sex ratio) was the highest in Punjab and Haryana (along with Gujarat and Maharashtra) which are not only among the richer regions of India, but also the regions where the share of population below the poverty line had reduced faster over 1972–3 to 1983. On the other hand, states like Uttar Pradesh and Bihar, with little change in poverty level, have shown insignificant increases in female agricultural labourers over the decade 1971–81 (Banerjee 1989a: WS15). The situation has remained more or less the same in 1991 except for Punjab and Uttar Pradesh.[26]

These discrepancies may result from a complex interplay of several factors all of which may not be comprehended very easily and systematically. Moreover, the intervening variables may be operating differently in different regions. For example, from the supply point of view, in the prosperous areas, the low level of female labour may be because of their socially induced withdrawal (forced or voluntary) from work in the public sphere with improvement in family income (Nagaraj 1989: 115; Nayyar 1989: 239), or it may be due to lack of demand in highly mechanised agriculture (with the average size of agricultural holdings going up) consequent upon the 'Green Revolution' in areas such as Haryana and Punjab. It is often argued that in such situations chemical fertilisers and introduction of weed-killers considerably narrows down the range of operations formerly available to women. Further, the introduction of labour-saving modern equipment indirectly affects the employment prospects of women by providing an unambiguous advantage to medium- and large-scale operational holdings which can no longer be sustained on family labour. The hired labour tends to be predominantly male (Bhalla 1989). Once again, the equation between 'Green Revolution' and decrease in female labour is not so simple. In the initial stages, the availability of greater irrigation facilities may generate additional demand for labour. It is in the second stage that the improved income effect of irrigation may lead to a withdrawal of the females from work resulting in a relative decline in female employment.

14

I am tempted to digress. Withdrawal from the public sphere does not necessarily imply decreasing workload. Only the nature of work changes and it may now be confined to the home. The Indian data show that unlike the pattern of labour consumption in crop activities where an inverse relationship exists between female labour and increasing farm size (implying better income and standard of living), in dairying this association tends to follow an inverted 'U' shaped curve, i.e., after an initial decrease with farm size, the female labour increases. This is clearly contingent upon the livestock (for dairy) owned which increases with a household's economic status. Thus, when classified on the basis of assets owned and arranged in an ascending order in 1977–8, in the bottom decile, only 5 per cent of households owned dairy animals as against 82 and 74 per cent in the top two deciles (World Bank 1991: 317). Clearly, the increase in female labour in dairy work with income is due to the fact that the livestock-related tasks are primarily performed within the household premises and do not involve contact with the outside world.

Even though a deviation, the observation does make it imperative that different types of labour such as family labour and hired labour (permanent and casual) be distinguished. By acknowledging this it is possible to accommodate these two apparently contradictory situations, i.e. withdrawal in certain sections of labour and increase in another. In any case, these two processes are not mutually exclusive.

As the family income improves, it is the female family labour that withdraws from the public sphere. This together with enhanced demand for labour as envisaged in the initial stages of agricultural development may open up venues for hired labour, both male and female. A similar kind of association has been noted in Bangladesh where the adoption of high-yield technology in crop production in villages has not only led to more hired female labour, but also to an eightfold increase in the average number of days they were hired as compared to those villages where such technology was minimally accepted (Solaiman 1988; World Bank 1990). At a later stage, with the substitution of labour by machines, however, female labour can be dispensed with.

Supply and demand both affect females' labour-market experience. While trying to account for the increase in female agricultural labour in certain states, Banerjee in her study, mentioned earlier, does not see it as a supply-induced phenomenon because that would have affected females' relative earnings adversely over the period under observation i.e., 1972–3 to 1983 which is not the case. She further maintains that the increase is due to increased job opportunities for females (1989a: WS 15). In this context it is possible to argue that the demand for agricultural labour can be gender neutral and therefore one can be substituted with the other. However, in India, despite considerable differences in terms of absolute participation rates, areas with high percentages of male agricultural labourers are also areas with high percentages of female agricultural labourers (Chatterji 1984: 10; Duvvury 1989b: 91; also see, Office of the Registrar General and Census Commissioner 1991: Table 7). A note of caution is in order. While the coexistence of male/female agricultural labourers in some regions may be due to an agriculturally developed context generating demand for labourers in general, as has been argued, in another agriculturally backward context it may mean a process of increased pauperisation and casualisation expanding the supply of agricultural labourers: both male and female, although more so for females (Duvvury 1989b). Here is a classic example of similar, if not identical, patterns being created by altogether different processes which defy a universally applicable explanatory framework. Also, an element of class rather than gender alone can be seen entering the analysis of the labour market. This point is taken up later.

Perhaps an intriguing relationship emerging in the 1991 census is of relevance here: with a few exceptions, in rural India, the areas of increase for female labour are those where the male labour participation has declined (Office of the Registrar General and Census Commissioner 1991c: 14). It is rather difficult to appraise this phenomenon until we have the detailed industrial classification, but does it mean that in the face of increasing pauperisation and increasing unemployment among the males, as argued by some, female workers have to come forward to shoulder the responsibility of earning the family income? We would like to leave this question open.

Very few studies on linkage between poverty and female labour are available for urban areas. However, a recent study observes that in urban India as a whole and in the urban population of individual states, the proportion of poor among women is consistently higher than among the males. Further, a disproportionately higher concentration of females is found in the households headed by women. The case is the same in rural areas, but in the latter case the percentage of these females below the poverty line is no different from all rural households taken together. Presumably males of these households work elsewhere in urban areas and remit some of their income back home. Such a support system is not available for females in urban areas. Under these circumstances, poverty is more likely to push urban females into the labour market. That this indeed is the case is substantiated by some case studies whereby it is observed that among the urban poor the ratio of females to males at work is twice as large as in the general population (National Institute of Urban Affairs (NIUA) 1990: 29, 1991: 37; Banerjee 1992). Moreover, the increase in the number of female workers in poor households is higher as compared to others (Bapat and Crook 1988). However, increasingly the only employment options available for poor urban women are in the 'informal' sector. They usually enter the market in petty trades and services or highly exploitative home-based work.

A related issue which is not discussed here pertains to the wage differential between males and females which is often explained in terms of differential skill formation among males and females, although it may be convincingly argued (as done by Joan Mencher in the present volume) that the low wages paid to female workers is in keeping with inherent inequality and often misleading patriarchal vision of the male as the main source of family income. However, some of the problems faced by these workers are integral to female labour as a whole, as is revealed by the discussion that follows.

Occupational structure and labour

It should be remembered that labour force participation rates of urban females are indeed remarkably low in India. With the exception of Sri Lanka and Nepal, other South Asian countries have even lower rates.

The latest figures available for India come from the 1991 census, which records only 9.74 per cent of urban females to be in the workforce. However, this shows an increase of 1.43 per cent over the 1981 figures. By itself inconsequential, this increase is noteworthy because over the same decade male labour force participation in urban areas declined from 49.06 per cent to 48.95 per cent. The per annum growth in female workers works out to be 6.1 per cent as against 3.5 per cent for male workers. Thus, there were 178 female workers per 1,000 male workers in urban India in 1991 as against 139 in 1981 (NIUA 1991; Office of the Registrar General and Census Commissioner 1991c; Gulati 1993: 256–7).

Workforce rates are lower in urban areas primarily because of the type of occupational shift from agricultural to non-agricultural activities (Sinha 1972; Banerjee 1985; Mehta 1990). Besides, in urban areas the traditional division of labour ceases to operate, and the complementary relationship among family members is replaced by the competitive nature between individual units of labour. Moreover, part-time and seasonal work that is easily available in rural areas (and can be combined with domestic obligations) is absent in towns and cities, which creates a wide gap between the employment rates of rural and urban women, rather than the improved economic conditions of the latter group as has been suggested in a few sociological studies (D'Souza 1959, 1969).

In keeping with the rural scenario, technological changes here require the acquisition of new skills which the women do not possess (Gadgil 1965; Boserup 1970; Government of India 1974; Omvedt 1980; Kannappan 1985; National Commission 1988).[27]

First of all, it is the lack of education and appropriate skills that limits women's employment opportunities in towns and cities; second, even in the contemporary situation, child care and household responsibilities, made more difficult in the absence of kith and kin support, are seen as primarily women's domain. Their options are further limited, in an already job-scarce situation, by sociocultural constraints which still retain their hold in urban areas (Singh 1984; Singh and Kelles-Viitanen 1987; NIUA 1991; Raju

17

present volume). However, certain sectors of employment such as *bidi* (country cigarettes) making which can be carried out at home, or industries like construction which are traditionally considered as family enterprises do provide avenues for female labour.

Both female and male work participation rates tend to decline with the increasing size of urban settlements (Sinha 1972; Bhalla and Kundu 1982). Male participation rates, however, are higher in urban settlements of more than 100,000 population (Bhalla and Kundu 1982; see Raju in present volume). On the other hand, for female workers, the larger the city, the lower is the employment rate (Sinha 1972; Mitra *et al.* 1979). This is because in metropolitan cities, the development of non-household employment provides higher employment opportunities for males and suppresses the overall level of female work participation. Only recently this situation shows some change whereby the negative relationship between city size and female labour participation is not all that clear. This may be because the distinction between the 'formal' and 'informal' sector is becoming increasingly blurred and much of the 'formal' work in an urban setting is being done by piece-rate female workers on contract at home for a factory. This trend is visible in Bangladesh and Pakistan also (World Bank 1989, 1990). What is more, female workers in manufacturing (other than household) industries actually show a significant increase with the increase in size of cities (Banerjee 1985: 18–19, 1989a: WS 16–17; Mitra and Mukhopadhyay 1989: 523; also see, Raju in present volume). Part of the variation in the female labour force by city size can be explained by the nature of occupational structure available in the cities and the extent to which urban settlements provide activities that are 'appropriate' for women.[28]

As far as rural areas are concerned, our comments on the recent situation are constrained by the available information and once again we are confined to the Indian data. In India, the rural composition of the workforce is still dominated by the agricultural sector. According to the 1991 provisional census, 87 per cent of rural female workers are in the agricultural workforce. This percentage has remained almost the same over the last decade. The decline in male rural workers in agriculture is only slight, about 2 per cent, and they also form a sizable proportion of male workers, i.e., about 77 per cent in 1991. At individual state level, however, livestock, construction and household industries emerge as significant categories for female labour indicating the importance of diversification of the rural economy in view of the limited capacity of the agricultural sector to absorb/create further increase in female labour.

In Nepal also it is the agricultural sector which is dominant (see Acharya in this volume). As far as cottage industries are concerned, although about one-third of their total employment is provided by female workers, in traditional industries such as woollen products and cotton textiles women's contribution is about three-quarters of the total labour (Islam and Shrestha

1987). In rural Bangladesh, however, the cottage industries have been the major employer of female labour essentially because these industries are located at home where unlike agriculture the females can participate freely and yet not enter the public domain (Cain *et al.* 1979; Hossain 1987).

Demographic considerations

It has been a recurring theme in our discussion that whatever has been the structure and trend of female labour force participation, their roles are seen primarily as housewives and caretakers. Consequently, changes in the age composition of a given population become crucial in affecting female participation in economic activities. Education is an intervening variable. However, we do not propose a detailed exposition of the education–employment link, but note in passing the much discussed 'U'-shaped curve between the two in the South Asian context. That is, at both ends of educational attainment, female participation in the labour force tends to be high. In general at the lower end, social constraints are relatively more relaxed in the face of economic compulsions and prevailing (lower) caste customs. At the middle level more females are likely to be from middle and higher castes where societal taboos and restrictions are most severe. A very small segment at the highest rung of educational achievement are once again spared from societal biases. Some of the nuances of this association are captured by Kalpana Bardhan in the present volume. Our limited concern is with the demographic changes as it relates to the labour market.

The case of Kerala in southern India as the most literate state in the country with high female status is too well known to obviate a detailed discussion, but it may be pointed out that it is also a state where unemployment rates are the highest, particularly so for women. One of the reasons behind this is the fact that the existing labour market cannot accommodate the educated job-seekers. Yet, from our point of view what is more interesting to note is that in Kerala, unlike most states of India, between 1971 and 1981 the number of males and females has declined significantly in the lower age group 0–14 accompanied by an upward shift in the prime working age groups 15–24 and 25–9, the growth in the female population being over 30 per cent and almost 50 per cent respectively (Eapen 1992: 2181). In urban areas, this shift is even more striking for women. This growth without a corresponding expansion in the labour market has evidently resulted in a decline in female work participation over the decade. Another disturbing finding is regarding married women who seem to be indirectly discouraged from seeking jobs in an overall job-scarcity situation (Eapen 1992). In a north Indian urban study, it was found that in the sample 75 per cent of male workers were married, but among the female workers this percentage was only 54 per cent (Papola 1986). Given the fact that in India even among urban women only 9 per cent remain unmarried after the age of 19, the importance of marital status as a potential constraint in entering the labour market cannot be overlooked (Banerjee 1992). The limitations posed by marital status extend into childcare and care for the elderly. Thus, the proportion of children in the population i.e., child–woman ratio and the proportion of those who are outside the labour force for various reasons i.e., dependents assume importance, particularly in the South Asian context where formal social securities do not exist.

These variables can act both ways. A high child–woman ratio may hold back adult females from participating in the labour force, or it may actually push them to work for want of additional income. What gets substituted by what would depend upon the opportunity costs involved, the age structure of the children in the household and the nature of work available (Malathy 1993). A simple bivariate analysis between labour participation rate and child–woman ratio is, therefore, not enough to yield definite answers and there is a need to incorporate other variables such as different stages of life cycle and resource position of the households (Irfan 1981). In the same way, a high dependency ratio may be a liability as well as a source of support at home. This description is illustrative rather than exhaustive, but it demonstrates amply that it is not always some innate biological limitations that constrain female participation in the labour market. Instead they often operate under socially constructed notions of role models inflicted upon them.

The caste/class/gender overlap[29]

A social sanction towards female appearance in the public domain is still a significant factor operating in the South Asian context. However, such sanctions are neither applicable uniformly nor with the same intensity in every region and for every caste group. The relaxation of the social taboos for females belonging to lower castes are often explained in terms of their active participation in gainful activities. In fact, this relationship can be envisaged as somewhat circular, in that the active contribution of these females to the household economy results in a more liberal attitude towards them which in turn may create an environment conducive for females to go out and seek employment outside the familial domain. As a corollary of this social reality, the varying presence of scheduled castes and scheduled tribes has been suggested as one of the components to explain variation in female employment. However, the hypothesised association rarely holds up under statistical scrutiny (Gulati 1975a; Reddy 1985).

It is true that, with very few exceptions, the participation rate among scheduled caste females happens to be higher than that of non-scheduled caste females. However, an elaborate study of urban female work participation in Andhra Pradesh, Karnataka, Madhya Pradesh and Uttar Pradesh (Raju 1981), and other relatively recent studies in urban Madhya Pradesh

(Raju 1987), metropolitan cities of India (NIUA 1991) and rural Tamilnadu (Nagaraj 1989) point towards (with the exception of Uttar Pradesh in Raju's study, 1981), a statistically significant positive correlation between the two rates. The scheduled caste male and female workers also exhibit a similar type of association. Among the non-scheduled caste workers, however, such a spatial covariation is conspicuous by its absence.

The covariance of male and female labour force participation rates among scheduled castes is not difficult to understand in view of their place in the hierarchical structure of Indian society. Scheduled caste males do not differ drastically from the females in education or skill attainment. This together with the fact that lower caste females are relatively less constrained to operate in the public sphere substantiates the observation that labour absorption of these castes is restricted more by job availability than by gender constraints (Raju 1981, 1982, 1987).

From our earlier discussion it may be recalled that in India, areas with high percentages of male agricultural labourers are also areas of high percentages of female agricultural labourers. However, such a close correspondence is not evident for workers in general. This relation holds for 1991 also.[30] At the risk of generalisation, it may be maintained that, with a few exceptions, agricultural labourers are often drawn from scheduled (and other lower) castes who belong to the poorer segment of the population. In such a situation gender identity no longer acts as a constraint. Joan Mencher (in the present volume) is apt when she remarks: for landless labourers it is not a question of male vs female in a given locale, but whether the person is capable to do certain tasks.

In contrast, for females belonging to non-scheduled castes, the norms that influence the extent of their involvement in labour outside homes are related to their position in the caste hierarchy and social stratification. In general, among these castes exclusion of females from the labour force outside the familial domain has become a symbol of high social status. It is ironic that despite a relatively better off economic and social position, the life-worlds of these females contract as they attain adulthood and move up the hierarchical caste structure (Kala 1976; Sen 1988).[31] Their lack of education and appropriate skills together with a more restricted social regime define their sphere of action more sharply in a mutually exclusive way *vis-à-vis* their male counterparts.

The situation becomes somewhat complicated when it is observed that, as against their male counterparts with whom they bear no spatial covariation, participation rates among non-scheduled caste females covary most significantly with those of scheduled caste females. This combined with the close spatial covariation that exists among scheduled caste male and female workers hints at a very intriguing social reality: the (bulk of) non-scheduled caste female workers share more in common with the scheduled caste workers rather than non-scheduled caste male workers!

21

At a speculative level, what the association among female workers suggests is that when a region-specific socioeconomic context becomes conducive, females, irrespective of their hierarchical position in the traditional social order, behave as a cohesive body and respond to the demand for labour in a similar manner. This may appear to be a far-fetched argument, but once the attainment in education and skills of females in general and that of scheduled caste male workers in particular is considered, the levels do not seem to vary much. The societal restrictions are, moreover, less severe for scheduled caste women. Under these circumstances it is reasonably justified to assume that given the social constraints, the bulk of those non-scheduled caste females who are in the labour market, particularly from lower and middle economic rungs, are not only working for somewhat comparable reasons, but their activity spheres are conditioned by the same set of constraints as scheduled caste males and females. Within the activity sphere, however, the tasks they do are clearly defined.

In this context, it is relevant to note that the proportion of female workers is very low in non-household industry where some kind of professional skill is usually required (Joshi 1976; Mitra *et al.* 1979). The more recent trends may be somewhat promising in that in 1983–4, relative to males, the share of female workdays doubled in manufacturing, but these opportunities have been in low-paid and insecure jobs (Deshpande 1992). According to one study in Lucknow (India), women's share in supervisory and executive posts is only 3 per cent even where they constitute one-fifth of the workforce. Further, women workers are seen to be concentrated in jobs where chances of vertical mobility and promotions are virtually non-existent (Papola 1986). This phenomenon explains the seemingly anomalous situation outlined here.

Translated into objective reality, what these observations suggest is that in certain situations females, irrespective of their caste/class identities, are subjugated to a certain kind of treatment as females governed by the organising rules of market mechanics.

What is also becoming clear is that in dealing with issues related to work or otherwise, care has to be taken to evaluate if what is being observed is part of the larger socioeconomic context and regional social formations or whether gender-specific explanations are called for (Saradamoni 1987: WS 3).[32] Most studies on women often fail to isolate the context, exploitation, and oppression of women *qua* women from exploitation of the poor in general (Rudra 1989).[33]

The argument is not for subsuming the gender issue under other issues as interpreted by some (Desai and Krishnaraj 1989: 1676), but a caution against an *a priori* notion of gender as a category. This becomes all the more relevant in the South Asian situation where the societies are highly structured and the development of gender identities, if it occurs at all, is of a very different dimension. Without acquiescing to the structural inequalities based on gender in these societies, sometimes the relative deprivation of females can be better understood and appreciated by analysing the processes at work affecting males as well.[34] This would not only place a given situation in a wider perspective, but would also expose latent relationships between the sexes that often remain obscured when females are posited in isolation from the rest of society (Singh 1975; Libbee 1977; Chatterji 1984; Karlekar 1985; Rudra 1989; Raju 1991).

The regional context

Consequently and notwithstanding the fact that in the South Asian context, scholars have often taken disparate positions on the various aspects of female labour leading to a certain degree of confusion, we risk proposing that it is still possible to situate the apparently contrasting claims against the back-drop of an overall regional-social context within which different structural, class/caste and gender-specific locations at various scales can be examined.

The broad regional differences between the northern Indian plain and the states in the south in terms of societal attitudes towards females that keep on surfacing in studies at various scales are well known. Our earlier discussion on poverty and labour provided some more clues as to how the overall pattern of female labour participation reflects the adaptation of regional ethos even among the poor.

It is to be reiterated that despite absolute differences, the labour force participation rates for females among the scheduled castes and non-scheduled castes are highly correlated indicating a close spatial covariation. This should *not* be seen as a contradiction of our statement regarding a less restrictive social regime for scheduled caste females and their relatively higher participation rates. In a given situation, scheduled castes – either in relation to non-scheduled castes, or in absolute terms – need to work more. It is also true that relative to the non-scheduled caste females, they enjoy a more liberal and egalitarian lifestyle in terms of their participation in the public domain for reasons outlined earlier.[35] But, it is important to note that women belonging to lower castes do not always operate in complete con-textual isolation from the rest of society. Further, depending upon given regional cultural ethos and the role model set up by the higher castes, with improvement in economic conditions, some of the lower caste communities tend to withdraw their women from labour force participation in the public domain as indicative of their improved status, a process termed as 'Sanskriti-sation' by Srinivas (1956; Lynch 1969).[36] What these micro-patterns suggest then is that, instead of expecting a linear relationship between the proportion of the scheduled population and the proportion of female workers, we need to ask where these low-status workers are located, what are the activities in which they are engaged, what is the working pattern of other females, and why females of low status are working in one area and not in another.

A similar process is reported elsewhere in this volume whereby the upward economic mobility in the Indian Muslim households seems to have translated into a greater degree of confinement for women (see Chapter 7, this volume).[37] This phenomenon is not only typical of India. In Pakistan, a slight improvement in family status seems to have enforced stricter enforcement of purdah, a trend that has been termed as 'negative modernisation' by Pastner (1974).

This emulation can also work in a reverse direction. For example, it may be recalled that in a cross-state analysis, Uttar Pradesh in north India was the only state which exhibited no spatial covariation between non-scheduled and scheduled caste female workers. This, combined with unusually lower labour participation rates, suggests that in the first place few non-scheduled caste females work in Uttar Pradesh and when they do, their sphere of activity becomes distinctly different from that of scheduled caste female workers. Since Uttar Pradesh is a state with large Muslim minorities and lingering traces of *ashraf* or 'noble' Muslim élite culture, it is presumably possible to argue that the Muslim legacy involves the reinforcement of an already existing tendency in northern India of secluding high caste females from the public domain (Raju 1981). Similarly, in Pakistan where purdah is widely practised, it has been observed that even generally non-secluded women are greatly constrained in their employment options due to the general attitude in favour of women's seclusion (Mohiuddin 1980).

It is likely, on the other hand, that in socially less restrictive areas higher caste females adopt more liberal 'morals, rituals and beliefs' which in many respects are antithetical to their own (Kalia 1961; Raju 1987). Such a broad conformity to the dictum of a given regional cultural realm by females from different community and caste hierarchies has in fact been found to hold in a score of analyses (Papanek 1975; Trivedi 1976; Sopher 1980, 1983; Raju 1982, 1987; Horowitz and Kishwar 1984: 70–1; Singh 1984; Agarwal 1989: WS 53; Bardhan 1990; 476–7; Lateef 1990: 120).[38]

Withdrawal of females from work in the public domain due to upward economic mobility or the process of sanskritisation, or their becoming more confined to homes through the institution of purdah, or the relaxation of specific taboos indicate that females do not have existence in contextual isolation and their behaviours are moulded by socially constructed societal norms which time and again are mediated through space (Miller 1981; Agarwal 1983: 92; Ahmed 1985; Bardhan 1985; Standing 1985; Harriss and Watson 1987: 85; United Nations 1989: 296–7).

At this point let me recapitulate the discussion in this section and arrive at certain conclusions: a) in the South Asian context, mere presence of economic opportunities is inadequate in explaining the extent and nature of female participation in the labour force; b) while poverty and related attributes may enforce females to enter the labour market, the extent to which they can do so is constrained by their memberships of particular caste, community and religious groups; c) however, the extent to which caste assumes importance depends upon the class/caste overlap: in situations where this overlap is substantial, caste identities cease to be very important and vice versa; further, existing caste mores can be reinforced/modified by the presence of certain religious groups and/or communities; and d) notwithstanding localised specific details, an overall regional context is vital in explaining the broad regional variation in female workforce participation.

Admittedly simplified, the question still remains: gender relations in labour reflect complex interactions of ecological, economic, demographic, ideological circumstances; females have multiple memberships:caste, class embedded in region-specific social formations and structures etc., but within these identities is there a common thread linking them together as members of a particular sex *vis-à-vis* the other sex? Although empirical evidence about gender relations in labour markets has repeatedly shown the subordinate positions of females in terms of tasks, wages, workload, skill formation and several other attributes, there is another set of observations to establish that females may not invariably be in a deprived position and in some cases their deprivation may be a part of a larger deprived setting. (Chatterji 1984; Unni 1989; Visaria and Minhas 1991; Raju 1991). Given this, the question that follows is: can we think of a broad theoretical framework within which the gender relations in the labour market can be placed?

Following Walby's exposition, we submit that the concept of patriarchy is central for our understanding in this direction.[39] Traditional patriarchy saw society polarised between men and women as two undifferentiated categories: the former exploiting the latter. However, the foregoing discussion clearly establishes the shaky grounds on which such a supposition stands. Walby defines patriarchy as 'a system of social structures and practice in which men dominate, oppress and exploit women'. However, the 'use of the term social structure is important . . ., since it clearly implies rejection both of biological determinism, and the notion that every individual man is in a dominant position and every woman in a subordinate one'. According to her, patriarchy has to be conceptualised at different levels of abstraction. The most abstract form would exist as a system of social relations. At another less abstract level, among other attributes, patriarchy consists of a) the patriarchal mode of production and b) patriarchal relations in paid work (Walby 1990: 20).

According to Walby, the contemporary positive changes for females which appear to be making an inroad into the existing patriarchal order do not mean elimination of that order. It is only the degree and form of patriarchy which are changing. It can be seen that her observation can be applied to the South Asian context where the recent changes in the labour market in terms of female education, women's increasing participation in manufacturing or in non-conventional jobs are not completely independent of prevailing gender ideology (see Chapter 11).

Bringing in social structure into the matrix as done by Walby broadens the concept of patriarchy to accommodate certain aspects of the differentiated (class) position of females *vis-à-vis* males.

Similarly, Walby's distinction between two main forms of patriarchy, private and public, is relevant. The household remains a place where the prevailing patriarchal ideology expropriates division of labour among various members. In the public sphere there are a wide range of intervening variables collectively operating in appropriating female labour. Neither of the two is free from patriarchal hold, but the degree and the form patriarchy assumed would differ considerably.

Depending upon the social structure and the degree of flexibility patriarchy would allow given the ascribed caste, community and ethnic status or acquired educational and skill attainments, females can bargain with the existing patriarchal structure (Kandiyoti 1988). These bargains may be explicit or very subtle.

THE VISIBLE AND THE INVISIBLE[40]

As seen, in examining some of the conventionally accepted propositions in the existing literature regarding regional variation in female labour participation, we are faced with recurring references to an overall regional context remaining vital. At the same time, there exists enough research to point out locational specificities which defy broad generalisations. Neither of these observational paradigms can be ignored. In the thematic organisation of our book, therefore, we have adopted an analytical framework at three levels: a) broad regional analysis with micro-level data from nationally published sources; b) sub-regional studies; and c) micro-level field observations. This scheme is adopted so that contextually specific sub-regional and local agro-ecological and/or techno-environmental demographic and social constraints which affect gender relations in the labour market can be placed against the backdrop provided by the overall regional perspective.

It needs to be pointed out that the bifurcation of the context is only for analytical purpose. The overall, sub-regional and local constraints interact in permutation and combination and not mutually exclusively. However, for analysing the female labour force, one needs to carefully scrutinise the localised difference as well as the broad similarities. It is possible that what is being propagated as a localised and contextually unique phenomenon turns out to be only a microcosm of the larger sociocultural context within which that micro-setting is located. Similarly, the broader observation may well be a collective articulation of what is happening at the micro-level. Analysing the issues in an overall context as well as at different scales bringing out the subtle nuances enables one not only to test propositions back and forth at both levels, but also helps one resist the tendency of finding simplistic explanations for what appears on the surface. Once this stance is accepted, it is possible to understand the contradictions in the literature and resolve seemingly conflicting views regarding female labour.

The complexities with which the seemingly mundane task of recording of female work is saddled with are evident now. Explaining variations therein and coming up with an explanatory framework that takes into account region-specific sociocultural, economic and structural constraints including caste/class/religious and gender overlaps is a much more uphill venture. Moreover, regional patterns of the workforce are to be interlinked with wider structures and processes in a society at large. The first paper in this collection by Kalpana Bardhan aims at this unenviable job whereby four sets of economic-structural-demographic and religious-cultural factors are analysed. They are a) income and assets at household level, b) education and skills of women, c) child–woman ratio and age of women, and d) level of economic growth along with region-specific presence or absence of social stigma associated with work outside the familial domain in the public sphere.

While exploring the interlinkages between female work and these factors, an attempt has also been made to understand and appreciate some of the apparently contradictory findings of many of the analyses regarding women in South Asia. In general, poor and illiterate women from economically and environmentally deprived regions seem to be working the most as are the women from agricultural regions as compared to those from urban-industrial contexts. There exists a 'U'-shaped curve between the female work participation rate and female education and family income. As has been documented elsewhere in the world, women outnumber men in casual, informal and insecure sectors with lower wages, whether agricultural or non-agricultural pursuits.

The main religious or caste-based cultural difference affecting female work participation is the stigma attached to women's visibility outside the home. Interestingly, this phenomenon has a distinct regional dimension whereby the northern region of South Asia (with the exception of Nepal) emerges as more rigid compared to the southern region.

This regional context becomes so pervasive argues Saraswati Raju that instead of a relatively homogeneous pattern and a distinct urban ambience largely independent of regional constraints, the pattern of female labour participation in cities follows that for the non-city and total population of the region in which they are situated. Moreover, females, irrespective of their position in the traditional caste hierarchy, follow the region-specific norms, biases and constraints in terms of their participation in labour outside the familial domain. However, when the gender relations are ana-lysed within the broader framework of regional differences, several nuances in the existing norms and restrictions surface differently for different caste groups.

In general, the household and the non-household manufacturing sectors are mutually exclusive and negatively related in terms of their absorption of female labour. The former sector seems to favour female work participation implying their concentration in home-based, unorganised and informal activities. However, cities in India, especially those with a million plus population, are no longer as hostile to women's work as before and an increase in the proportion of female workers in the respective population is associated with an increase in their share in the non-household sector. Additionally, of late, the household and the non-household activities seem to have become complementary.

In the metropolitan centres and large cities, the occupational structures available seem to have a far more significant bearing upon female labour force participation as compared to small and intermediate cities where the religious composition of the population plays a vital role.

Joan Mencher's study of India's rice-growing states of West Bengal, Tamil Nadu and Kerala reveals variations in the gender division of labour from state to state (macro), village to village (micro) and caste to community. Mencher identifies four relevant facts about women's work in the subcontinent's rice ecology:

a) women do a large part of the heavy manual work notwithstanding prevailing concepts that women's work is easier than men's;
b) clear job segregation exists on the fields by way of male tasks and female tasks, though micro and macro variations in the nature of tasks performed often point to techno-environmental or historical-political factors;
c) women's income is a major source of the family income; and
d) manual work is generally considered degrading by the upper caste and class groups. Upward household mobility, however, may often mean increased land size, production and work volume resulting in women carrying a larger than usual workload.

Women continue to play a major role, putting in as much as 407 female work hours per acre per season against 106 hours of work by the males, despite the dramatic changes that have occurred in agrarian relations in the modernised rice economy. Gender wage differentials are pervasive; wage laws continue to be discriminatory based on the common notion of lower productivity rates of women. Women are barred from formally taking on managerial and supervisory roles although they might informally be in full charge of farm operations at home. Mencher discusses whether or not lack of control over the plough translates into lack of control over products of labour and assets.

The first three chapters outlining the general structure of the workforce set a stage against which some of the issues related to female labour can be examined. As discussed earlier, the major issues still revolve round the definition of 'work' which, as it stands now, is extremely limiting. Even outside the market economy, the so-called 'household work' contains quite

28

a few components which can be easily substituted for wages. The chapters in Part II bear testimony to this ground reality.

A word of explanation, however, is in order. While objecting to the artificial division of labour in household and non-household categories for reasons outlined earlier, we have still retained it as the organising principle of this section. This is essentially for want of a better alternative and is a reflection on the existing system of data collection.

In her analysis of the South Asian region (as a whole) in Chapter 1, in terms of the inside/outside dichotomy of work and women's visibility in the public domain, Bardhan sees Nepal as an exception in the otherwise rigid northern South Asian region. However, when the scale of analysis moves from the macro perspective to meso level observation as has been done by Meena Acharya in Chapter 4, two distinct groups in Nepal – the Tibeto-Burman and the Indo-Aryan in mountain and hills, and *Tarai* regions respectively – emerge differently in that there is a strict seclusion of women and prevalence of purdah among the women of the latter group. Notwithstanding this, the subsistence nature of the economy and mass poverty determine the work patterns and life options of women in Nepal. According to her, decisions about female labour supply have two components: decision to participate in the labour market and the number of work hours to be spent. The economic, caste and ethnic statuses operate more significantly at the entry point only. Once in the labour pool, demand and demographic factors determine the work hours. Thus, cultural ideology and economic necessity combine to produce the regional variation in female labour participation in Nepal.

Bagchi's study of women's work in central India uses the concept of *Chakra* from the classical Indian language of Sanskrit to define in a circular graphic form the inter-relationships between domestic and extra-domestic household work that is economic in nature by being either income generating or income saving. Domestic fuel collection is presented as an example of such a household activity, whereas women's farm-related work is described as extra-domestic. Both these categories of work are crucial to the family sustenance; women juggle time and space to accommodate the 'economic' tasks with 'service' chores of washing, cleaning, cooking, feeding and tending. Both these activity groups suffer from invisibility, immeasurability and undervaluation.

Jacobson's photographic analysis of a single central Indian village drives home the point that the core of women's work is *Chula-chakki* (food processing) which unavoidably includes provision of water, fuel and fodder. This responsibility not only involves long hours and arduous work but also low levels of technology. It does not, however, deter the workers from crafting their own stoves, processing their own fuel cakes or doing their own roofing, plastering and whitewashing at their dwelling sites. Women's out-door work is conspicuous and essential yet remains invisible to the discerning eye. Women are culturally barred from performing heavy work, e.g.

29

driving the plough or the bullock cart, yet they are seen all over the countryside crouching, bending, kneeling, carrying heavy headloads, but always in a subordinate state supervised by their male counterparts. Equal opportunities of education, meager as they are, might be the only hope of bringing about a change in the social evaluation of women's roles as has been witnessed by a selective band of high-profile women workers in non-conventional fields.

Huma Ahmed-Ghosh's study of a north Indian village provides a clear and straightforward discussion of the varying connotations of work outside the home for women of different castes and socioeconomic groups and uncovers the interaction of gender and class. In doing so, she analyses the processes of readjustment of the female workforce following the relative success of the Green Revolution which produced diversification of farm operations for the landed households and creation of off-farm jobs with fixed wages and benefits for landless households. The impact on women of this upward mobility was uneven across caste and community groups. Ghosh's research describes withdrawal of women of scheduled castes from agricultural wage work with improvement in families' financial status through fixed wages and benefits. Women of the upper castes and class, on the other hand, found themselves saddled with new responsibilities as their families diversified their operations and as hired labour becomes scarce in the village. Only the women of the Muslim community were unaffected by this dynamism.

Gender division of labour, women's work and the impact of a new highway linking the remote community of the Karakorum mountains administered by Pakistan is the focus of Sabine Felmy's study in Chapter 8. The Hunzucuk of the Hunza Valley are a close-knit agricultural-pastoral community living in close harmony with an inhospitable mountain environment. The community is distinguishable from the rest of Pakistan in its affiliation with the progressive Ismaili faith started in early 1900 by His Highness the Aga Khan, and by the absence of purdah among women.

Despite the prevailing traditional gender roles – male children are trained to manage outside affairs while female children are trained to manage domestic affairs – a clear-cut differentiation has not traditionally existed in the Hunza household. Men and women have been exchanging roles as and when required making it easier for women to take on men's work in recent years when the latter began to opt in large numbers for non-agrarian jobs made available by the opening of the new highway. Such a readjustment of the labour force has albeit forced women into longer work hours though still allowing them the option to hire labour and the choice of abstaining from hard manual chores like ploughing, pruning, planting, etc. A dynamic approach to gender division of labour and roles of women has a positive impact on integration of women in modernisation schemes by allowing free input of women in decision making, a case which might stand out in sharp contrast to development experiences elsewhere in Pakistan or in South Asia where development approaches have largely been gender blind.

A recurring theme in this collection has been about the gender ideologies as constructed by prevailing patriarchal order and influenced by specific social structure. However, in some cases females have been able to bargain with patriarchy and in the process acquired a certain degree of autonomy. One would assume this process to take place with modernisation, but the question that one needs to ask is: how stable or fragile is the acquisition of autonomy? Schrijvers' study of female labour of the North-Central Province of Sri Lanka provides an answer. She observes that a shift from an ancient system of swidden agriculture (*chena*), by which the main sustenance crop of millet was cultivated, to the modern commercial paddy agriculture reduced the key role played by women in the earlier system to a more subordinate role in the latter.

A parallel impact of the subsequent economic transformation has been loss of women's autonomy and self-esteem, being forced into wage labour by the new system and thus encountering economic and biological sub-ordination to males within the household. Schrijvers explains the various elements of women's subordination as part of a continuous interconnected process. In her research, she has adopted the participatory/action research model instead of passive observation so that women could be helped to regain their lost autonomy and self-esteem. The women's farm was begun with this purpose in mind and worked splendidly in reaching the objectives until the political instability and violence of the island turned the clock backwards.

Interestingly, Chapter 10 by Joyce Aschenbrenner is a reinterpretation of data generated twenty-five years ago on conditions of women's lives and work in a Punjabi village in northern Pakistan in the light of subsequent scholarship. The premise here is of a continuity, a perenniality in the male–female power relationship in traditional village structure, and the presumption that operating within such a structure, women are able to, more often than not, manoeuvre it to their advantage. Aschenbrenner's view is shared by many scholars in that the power women wield in a traditional village system rests to a great extent on their participation in solidarity groups – working together, participating in ritual, recreational activities, sharing experiences and maintaining loyalties with other women.

This view is reinforced by the results of a more recent village-level income-generation project for women in Shahkot village in the same province where a successful hand-embroidery enterprise for women was able to demonstrate the value of group interaction combined with income-generation work in raising awareness and self-esteem, thereby reducing dependence on men (Khan *et al.* 1989). Also available are in-stances from elsewhere in the subcontinent of women creating their own solidarity groups (*Mahila Mandals*) to help them come together in sharing, articulating and voicing their experiences, needs and aspirations. Aschenbrenner's point is shared by the editors that improvement in people's lives should be based on the strengths displayed by members in everyday lives rather than on the weaknesses of the community at large. Micro-level village studies of this nature are invaluable in exposing such realities.

Chapter 11 attempts to sum up the various expositions put forward by the authors in the preceding chapters. The gaps in the existing research are identified and some directions for future research are suggested particularly now that the South Asian countries are undergoing major changes in the economy.

NOTES

1 Data for all countries except India relate to the economically active population aged 15 and over for the year 1990. The comparable age-specific data are not yet available for India and the percentage refers to the rate for 1991. However, in 1981 the comparable percentage was 20.

2 The International conference on appropriate agricultural technology for farm women organised by the Indian Council of Agricultural Research in New Delhi in collaboration with the Rice Research Institute, Manila, Philippines in 1988 brought out very elaborate accounts of a wide range of these activities by women including seed-selection, seed-treatment, processing, storage of grains in agriculture, commercial and otherwise, fisheries, marine and fresh water, sericulture, and pisciculture, etc. Also, see Devadas *et al.* (1988).

3 These activities were: maintenance of kitchen garden, orchards, etc, work in household poultry and dairy, free collection of fish, small game, etc. free collection of firewood, husking paddy, preparation of jaggery, grinding of foodgrains, preparation of cow-dung cakes to be used as fuel, sewing and tailoring, tutoring of children, bringing water from outside the household premises and outside the villages.

4 The Thirteenth International Conference (1982) on Labour Statistics by UN/ILO resolved that work for pay or in anticipation of profit must include 'all production and processing of primary products, fixed assets and other commodities whether for market, for barter or for own consumption' (Anker 1983: 713; Jose 1989: 3–4). The current version of the United Nations Systems of National Accounts recommends wider coverage of non-monetary as well as monetary goods and services in the concept of economic activity. However, goods and services produced in the household for its own consumption are only covered if those goods are also marketed (United Nations 1991).

5 In West Bengal the peak household work coincides with harvesting of *Aman* and sowing of *Boro*. In Rajasthan also, a similar relationship exists. This is because while harvesting and allied income activities are at the peak, the women of landed peasants are busy at home with processing of grain, and in cooking, serving and washing chores required to feed farm hands (Jain 1985: 230–1). Asuri and Mahadevappa report that in Karnataka's sericulture, in times of intensive work load on the farm, the farm hands are fed at least twice. Preparing food for these workers becomes the top priority for the women of landed households (Asuri and Mahadevappa 1988: 6).

6 During the field work in a village in 1988 near an urban centre in Uttar Pradesh,

India, this phenomenon was observed by Raju in its intensity. The scheduled caste men, who customarily used to clean cattle-sheds and attend to various chores associated with livestock care had found government jobs in the nearby city (spurred by a job-reservation policy of the government for this deprived section). This necessitated women from well-off families undertaking their work, and the women resented it.

7 Duvvury and Issac have listed various activities in which women engage themselves dependent on their proximity to the household ranging from wage (and salaried) employment as the most distant (from home) and the most visible, to domestic work which is least visible (1989: 1). Kala has the specifics about the distances allowed and the limits imposed on women for agricultural tasks before they become 'outside' work and hence not meant for women (Kala 1976). Also, see Mies (1981).

8 Data from village studies in Pakistan show that farm activities (at times regarded as an extension of the women's household jobs) increase tremendously during the harvesting seasons (Saeed 1966). A micro study from Bangladesh reports that apart from preparing the puffed, perched and popped rice consumed by the local people, the women also supervise over the storage, the turnover of the year's supply of foodgrains and seeds to be used for next year's planting so that good germination can take place (Abdullah and Zeidenstein 1975, 1982). Incidentally, women are traditionally known for their ability for seed selection (Swaminathan 1985). Women's contribution in post-harvest operations related to paddy, either for self-consumption or for business purposes is brought out in yet another study of villages in West Bengal, India which reports that there in all, 55.29 'man' days of work/person/year are available in post-harvest activities. Out of these 44.36 'man' days (80.23 per cent) are actually women days spent in parboiling of paddy, separating and making of puffed rice (Mitra, et al. 1988b: 3, 16; also, see Note 1 in this chapter).

9 Time allocation studies have their limitations as well. Despite great emphasis on observation, ultimately it is on the basis of recall that data are usually gathered. See Krishnamurthy (1985: 253–5).

10 For details see Unni (1989), Nayyar (1987), Kundu and Premi (1992) and Patel (1989)

11 This 'slack period' varies significantly with season and region. For example, Kumari and Chari observe that in tribal areas of Andhra Pradesh, the women have longest hours of work in the *kharif* season. Whereas the women in areas of northern light soils and black soil tracts put more hours of work in the *rabi* season. In the areas of southern light soils, however, both in *kharif* and *rabi* seasons women are overloaded with work in farm and off-farm operations (Kumari and Chari 1988: 8). Also, see Radheyshyam *et al.* (1988) for their account of women's involvement in agricultural plus non-agricultural work and how they still remain outside the 'principal activity' criteria.

12 A similar phenomenon has been observed in the case of Peruvian census data where initial sorting of persons into workers and non-workers on the basis of 'principal activity' led to a substantial reduction of female workers in 1971 (Deere and Leal 1982).

13 Census authorities do not attach much importance to this additional instruction to enumerators because according to them this clause has always been there even if not explicitly stated. They seem to attach far more importance to the societal biases, both of enumerators and respondents in identifying and reporting women's work (personal communication). However, the impact of this rider on recording the actual rates of female work participation together with media-

33

oriented public awareness campaigns on hitherto less visible activities by women seem to have helped as the provisional figures on women's work in 1991 do show a sharper increase in female labour participation rates as compared to males. However, as pointed out by Visaria (1991), the increase *per se* should not be a matter of jubilation. The structure and composition of female labour and the nature of the increase and shift in various sectors must be carefully scrutinised. See Krishnaraj (1990) for a detailed account of various recommendations by UNIFEM on changes in the data collection format on women's work for the 1991 census and the final version adopted by the authorities.

14 In this context, the social construction of gender, whereby certain tasks and activities are ascribed to women, which are not related to their biological roles, is important as it imposes structural constraints on their spatial behaviour.

15 This situation is gradually changing. According to a recent report of six Indian cities by the National Institute of Urban Affairs (NIUA 1990), most of the females who were economically active perceived themselves as workers. Those who did not do so were essentially engaged in livestock rearing where some of the products obtained were for home consumption and the selling of the rest was done by males.

16 In 'Food for Work' programmes in parts of South Asia, most of the women seeking work were those who had dependents with little or no support from their male counterparts. See Agarwal (1985).

17 A study on vendors in Madras outlines the norms of female seclusion and its impact on the trading performance of women (Lessinger 1985). See Bardhan (1985: footnote 28) and Andrea Singh (1978: 77–9).

18 Following the exposition put forward by Nagaraj (1989), we feel that it is very difficult to have a clear-cut distinction between supply and demand factors as they operate in conjunction with each other to reflect basic socioeconomic processes at work.

19 According to Beneria and Sen, the empirical insights appear to support a theoretical model of fragmented labour markets rather than a model of a competitive labour market, but the categories are not neoclassical. What appears to Boserup to be a technically determined correlation between plough cultivation and women's lower participation in field work has its roots in the social relations of production and reproduction (Beneria and Sen 1988: 361).

20 In general, women do very little of ploughing and engage in transplanting, weeding and harvesting. However, this differs and in Tamilnadu, women provide more than 70 per cent of total labour in transplanting and weeding whereas only 25 per cent of transplanting is done by them in West Bengal. Even within a state there exists community and region-wise variations in societal approval of tasks for women. See Chapter 3.

21 According to Miller, an important contrast has to be drawn between rice cultivation through broadcasting seeds and by transplanting seedlings. The former method is more labour intensive as compared to the latter.

22 If the region-specific culture explanation is to be accepted, this lining up of Orissa with the wheat-growing north needs no elaboration for Orissa is located in the northern social space where traditionally there are more restrictions on women's participation in the public domain. It is a case of sociocultural mores overriding ecologically induced patterns (Government of India 1974; Agarwal 1986b; Miller 1981, 1989).

23 The author is grateful to ILO, Geneva for permitting the use of copyright material in Ruchira Chatterji's paper.

24 This association may, however, be operating in the reverse direction, i.e. the

higher expenditure pattern may in fact be related with more earning members in the same household.

25 In a recent study by Acharya and Mathrani (1993), the variation in inter-state labour participation rates of rural females is explained in terms of the varying nature of land systems. Accordingly, in eastern and northern parts of India the land has historically been under the *Zamindari* (landlord) system in which tillers received low returns and low social status. The lingering impact of an association between physical labour and low status has been such that the tendency of withdrawing females from physical labour as soon as family conditions would permit remains. In contrast, the southern and western states had the *Ryotwari* land system in which land had been under the control of middle castes who directly worked on the land. Also, see Acharya (1987).

26 In Punjab, the arrest in the growth of female agricultural labourers may be viewed as a part of overall declining agricultural activities consequent upon growing terrorism in the past few years. In Uttar Pradesh, on the other hand, the 1991 census has registered an unprecedented increase in the overall female labour force participation rate which seems to have raised the sex ratio of agricultural labourers as well.

27 This situation where the women are allocated low-skilled jobs on account of low attainment in modern skills may vary. The Marga survey in Sri Lanka in early 1978 reveals that while higher qualifications and modern skills are prerequisites for obtaining jobs in urban areas, they do not necessarily help in achieving that goal. See Ahooja-Patel (1986: 229–30). A similar observation has been made by Acharya and Mathrani (1993: 44–5).

28 See Sopher (1980: ch. 4, footnote 108) and Singh (1984).

29 Our analysis in this section relates essentially to the Indian situation, which is further limited by the fact that the caste-wise data for the country as a whole is available in the census only, which divides the population into two segments: the scheduled castes and the rest, which may be loosely termed as 'higher castes'. However, these high castes consist of varying groups living under a wide range of social and economic conditions, but the data available (such as the one we are interested in) do not allow such distinctions to be made. The term 'scheduled' refers to specification on a constitutional schedule of castes deemed to have been historically the most backward and socially deprived segment of the Indian population *vis-à-vis* other castes. The list remains substantially the same as originally prepared by the British in 1935.

30 These calculations are done on the basis of the Office of the Registrar General and Census Commissioner, Paper 3 of 1991 census, India.

31 This became clear in several village surveys conducted by the Centre for the Study of Regional Development, Jawaharlal Nehru University whereby the scheduled caste females displayed a much greater spatial awareness in and around the villages as compared to those belonging to higher castes and Muslim communities.

32 As pointed out by K. Bardhan, sometimes both economic (income/landholding etc.) and social (caste/ethnic) factors have a bearing upon the rate and pattern of female labour participation with the latter overriding the former as 'the difference by caste/ethnic divisions within a farm-size group is stronger than the difference across farm-size groups within a caste/ethnic category' (Bardhan 1983: 59).

33 Stoler in her study of class structure and female autonomy in rural Java puts it succinctly, and we quote:

> We cannot . . . view women as a homogeneous group . . . nor can we assume that exploitation will occur primarily along sexual lines. Changes in the

structure of precolonial, colonial and present Javanese rural economy did not catalyze an increased dichotomy of sexual roles, but rather an increased scarcity and concentration of strategic resources. These changes adversely affected both men and women in the lower strata of village society. . . . Poverty is indeed shared but only among the already impoverished men and women.

(Stoler 1977a: 88)

34 It is frequently argued that the process of development has lowered the status of women in general and in South Asia in particular by reducing their participation in the labour force. There are contrasting evidences (Sen, G. 1983; Duvvury 1989a). A recent comprehensive study on changes in women's employment denies the alleged decline in the female labour force (Unni 1989). One may dismiss this debate on the decline/non-decline of the female labour force as inconclusive, but the more relevant aspect of labour force participation that is of interest is when Swaminathan argues that the trend of declining labour participation by females is not confined only to them, the underprivileged as a group is subject to a similar fate (Swaminathan 1975). Gulati's study of unemployment among female agricultural labourers in India shows that though there are wide inter-state differences in the levels of unemployment, male and female levels covary significantly, i.e., female unemployment is higher in states where male unemployment is also high and vice versa (Gulati 1976).

35 Horowitz and Kishwar observe how in a household of agricultural labourers (who almost invariably belong to low castes), the male members seem to help the women much more with domestic tasks as compared to high caste Jats (Horowitz and Kishwar 1984: 90; Sen 1988).

36 The term 'Sanskritisation' was first used by Srinivas in his book *Religion and Society among the Coorgs of South India* (1952: 30). It means adaptation by lower castes to the ways of life such as vegetarianism, teetotalism, rituals, customs and rites, etc. of higher castes, especially Brahmans. By doing so, it is possible for lower castes, in a generation or two, to rise to a higher position in the traditional caste hierarchy.

37 However, in doing so, the regional Muslim community is more influenced by local norms of social behaviour than by other Muslim communities. And, the Muslim legacy is only reinforcing the already existing tendency of secluding high-caste women from the public domain (Lateef 1990: 121; Raju 1981: 19).

38 If, indeed, we are to view the extent of female participation in the labour force as a function of social ethos, then the existing social behaviour towards women should influence both the urban and rural communities equally of a given region. In such a situation one may expect both the urban and rural levels of female labour participation to covary spatially indicating conformation to broad regional culture. In fact, in India, despite the large differences between rural and urban conditions, a common geographic pattern of variance does exist (Raju 1982, 1987; Bardhan 1985). Our thesis is substantiated by yet another study where the significantly close coefficient of variation between the urban and rural female labour participation rates led the authors to suggest that there exists 'a commonality in the social forces operating in both settings through the extent to which different social forces operate in each would be dissimilar' (Acharya and Mathrani 1993: 47). Also, see Sundaram (1988). Such a correspondence is found even in variables like literacy which is generally seen as an urban attribute (Sopher 1980).

39 This section draws heavily from Sylvia Walby (1990).

40 The author wishes to acknowledge with gratitude the help she received from Deipica Bagchi in writing this section.

Part I
THE STRUCTURE OF THE WORKFORCE

1

WORK IN SOUTH ASIA
An inter-regional perspective

Kalpana Bardhan[1]

Much has been studied and written in the past two decades about how the processes of industrial and agricultural development in South Asia have affected women's work and occupational pattern and how these effects have impinged on their well-being and their position in family and society. Often, questions are raised and claims made, generalising from either one-country macro data or a few localised case studies. Since there are large inter-regional differences within South Asia both in the experience of development and in the way it has affected women in their work and work-related life conditions, it is important to look into the regional differences that may have explanatory significance in addition to the average situation and trend.

Often, too, despite the sincerely feminist concerns of many studies, a degree of frustrating confusion has been produced by their seeming to point to conclusions that either contradict each other or stand about without having their connections sorted out, their apparent conflicts explained or resolved, and their real conflicts or tradeoffs identified and assessed. The problems of women's work and work-related conditions of life are viewed differently by various studies. On several issues, they seem to be saying opposite things, e.g. women have too much work versus they have greater unemployment and underemployment; a decrease in economic activity rate indicates worsening status of women versus an increase in economic activity rate reflects distress labour under the pressure of poverty and uncontrolled exploitation; women are being displaced from gainful employment versus they are being hired heavily in certain job sectors for their easier exploitability; housework being undervalued and unrecognised versus housework being a marginalising factor for the working poor; family being the site of oppression versus the disintegration of family precipitating female and child destitution and ill-being; the state adversely affecting women through growth-oriented policies versus the state adversely affecting women through stagnation-fostering policies; laws purporting to help working women end up hurting them versus laws should be enacted to curb the exploitation of women and their displacement from jobs, and to reserve job quotas in government services and public sectors.

39

Undoubtedly, there are elements of truth in these opposite claims in the stratified societies of South Asia. But either because the different claims hold for different sectors of women, or because they are the flip sides, or perhaps sequential effects, of a certain conjoint process, it is important to lay out how the different conclusions relate to different regions, different structural contexts and different sections of the female population. The seemingly opposite conclusions reached by different studies on women's work may be resolved by examining the structures of their regional *and* class location. Within South Asia, there are major inter-country differences in a given point of time as well as in the way things have changed or not changed over time. There are major differences among sub-regions within a country and among women from the different socioeconomic groups within each country. By trying to identify and understand these differences, we can see in perspective the apparently opposite conclusions that have been reached in many studies on women and the different kinds of work they do. That of course, is a huge task – too huge for one chapter.

This chapter is about the similarities and the differences among the South Asian countries and sub-regions in respect of five points: a) the overall and age-wise rates of economic activities of women, b) the occupational distribution of rural and urban women workers, c) their educational access and attainment levels, which is perhaps the most crucial determinant of occupational mobility, d) the effect of environment (access to water, forests or commons) on women's work among the rural land-poor in particular, and e) the participation of the female working poor in organised efforts to ameliorate their working and living conditions, in forging collective institutions of self-employed or piece-wage home-based workers in order to upgrade work conditions and returns in the low-end occupations they are in – upgrade in small but significant ways, even if without occupational mobility, by lessening exploitation and generating voice and hope.

The reasons for comparing the last three factors on a regional basis are briefly these. Education is one of the main correlates of women's participation in, or access to, non-agricultural employment in the formal sectors: what promotes girls' schooling in a period shows up in occupational advancement ten or twenty years later; what hinders it also shows up in the persistent sex-gap in the quality and quantity of work. Reduced access to the commons and the forests for subsistence resources overloads the land-poor rural women's energy expenditure on essential subsistence tasks. Ameliorative organisation of low-income female occupations, the experience of which has varied inter-regionally, captures important sociopolitical aspects of regional variations in women's work conditions not explained by the usual set of economic, agroclimatic and natural-resource variables considered in most inter-regional analyses.

The comparisons attempted here are both among the South Asian countries, and among the sub-regions, mostly inter-state, within India and

Pakistan, as data permit. This is to utilise the available data to the fullest possible extent, and to compensate data deficiencies across the South Asian countries with inter-regional comparisons within India, a country of continental diversity and with by far the most comprehensive data base. Besides, as noted in the following sections, some of the patterns observed in inter-country variations are paralleled by those observed in inter-state variations within India, and to an extent also Pakistan.

The first section examines the inter-country variations in the rates of economic activities of the female population, and in the female proportion of the workforce in 1960 and 1980, and some of the factors that explain these variations. The second section is on the pattern of inter-state variations in these two indicators within India. The third section examines the relation between education and rates of economic activities of women workers. The fourth section deals with the age-wise pattern of the rate of economic activities of women in the five countries in 1960 and 1980. The fifth section is on the women workers' employment pattern or occupational distribution. The sixth section concentrates on the various experiences of organisation of the female working poor in efforts to ameliorate their work and living conditions through self-help and lobbying.

INTER-COUNTRY AND TEMPORAL RATES OF FEMALE LABOUR PARTICIPATION AND WORKER RATIO

It is now generally recognised that the national census data, the source most used in international comparisons, seriously underestimates a) the extent of women's participation in agricultural and home-based work, and b) the predominantly part-time female workforce. If the causes of such under-enumeration were invariant over time and across regions, countries and cultures, then although the absolute figures would be suspect, the pattern of variations shown would be reasonably reliable. So would be the structural relationship of that pattern with the variations in economic and social indicators. Of the causes of under-enumeration of women's work, the one that varies greatly across the South Asian regions is the religious-cultural stigma attached to women doing farm work and marketing. This stigma, and the associated under-reporting of female farm work is generally greater in Bangladesh and Pakistan than in Nepal, Sri Lanka and India, in that order. Within India it tends to be greater in the north and the east than in the south and the west. However, a rise in household income, availability of opportunities for education and alternative employment tend to induce the withdrawal of females from work in the field and wage-labour irrespective of a strong presence or an absence of this stigma. Thus, the way in which the pattern of regional variations tends to behave over time is likely to reflect the operation of certain economic and structural factors in addition to or in spite of the effects of this cultural difference on women's work. As far as the temporal comparison is concerned, as with improvement in female education and employment opportunities which may vary regionally, even in census data, certain patterns of inter-regional variation in the female worker–population ratio are observed which correlate with the variation in economic and structural conditions. The significance of the structural-economic factors, notwithstanding the cultural differences, underscores the case both for regional dispersion of educational resources and economic growth resources and for targeting education and employment opportunities on the female poor, especially in rural areas and in favour of the younger age groups more likely to utilise an increased availability of better options.

The invisibility of women's home-based work and its integration with housework constitute another major cause of the census under-enumeration, both by investigators and by respondents, especially by male heads of household (Abdullah and Zeidenstein 1982; Acharya and Bennett 1981). The interregional variation due to this is related partly to the cultural factor discussed earlier but partly also to the inter regional variation in the nature and extent of women's home-based work in processing and expenditure saving for family use. This second factor tends to be negatively correlated with the household economic level and with the regional environment of economic opportunities. In other words, households poorer in income/asset level and located in an economically stagnant region tend to have larger proportions of female worktime indirectly spent in the income-stretching and expenditure-saving activities that tend to be under-reported the most in large-scale data generation such as the census. The time women spend in collecting subsistence material from the commons and in processing and producing things for household consumption tends to be greater in the more deprived regions and for women poorest in material and educational assets. It is this kind of non-wage distress work that goes unreported more often than any other work. Besides, the time spent in home-based self-employment or piece-rate wage work, which is less likely to go unreported, tends to be greater in rural regions that are either locally more developed or better linked with urban areas and among women with somewhat better assets or educational levels.

It would thus seem that regions or sub-regions with greater rural development (generating local sources of employment), or those with more access to extra-local sources of employment arising from urban or industrial development are likely to have better enumeration of the female worker–population ratio because of the greater prevalence of wage-employment and directly remunerative self-employment as against the work of gathering and processing at subsistence level. What this would lead one to expect in inter-regional (cum inter-temporal) patterns is that, given the extent of asset poverty in a rural region, greater agricultural development and urban growth linkages would generate higher female worker–population ratio even in the existing data. Less asset poverty in a region would also mean less of the kinds of work that go unreported.

A final problem in comparisons either over time for one country or across countries arises from differences in the criterion used in different national censuses for the number of days of reported work in a year that qualifies a person to be categorised as a worker. We know that part-time work is far more common among women than men, and that the home-based work most prone to under-enumeration is far greater in the case of women. As the census data on workers in most of the countries in question are based on around 180 days of work per year as the minimum cut-off point, a significant part of the actual variations in women's work between years and between countries tends to be left out, although the pattern of variation in the so-called main workers would be reliable. For long-term comparison from 1960 to 1980, a good way to look for supporting evidence for the structural correlates is to examine the regional variations within India. The Indian census data have been better adjusted for comparability (of main or main plus part-time workers), and the more comprehensive employment data from the National Sample Surveys are available for inter-state comparison between 1972–3 and 1983. This chapter's analysis of the regional perspective is based on both inter-country data and inter-state data within India covering the 1960s and the 1970s. There is indeed remarkable coherence and consistency in the results.

In spite of the well-known causes of underestimation in the labour force participation rate (LFPR) data for women, the inter-country and intra-country comparisons reveal a clear pattern that cannot be explained away by the limitations of data (Table 1.1). The broad pattern seems to be consistent with the operation of three explanatory factors: a) to what extent the region was or still is rural/agricultural; b) the extent of female engagement in agriculture, which in South Asia is strongly linked both to i) religious or ethnic cultural constraints, ii) the composition of crops, iii) the nature and extent of supplementary non-farm activities, and c) the female incidence of literacy, especially of primary and middle-level education along with the availability and growth of jobs that are semiskilled or require some school education. Across the five countries, and between 1960 and 1980 in these countries, the female share of the workforce and the female LFPR (items 1 and 5 in Table 1.1) move upward with greater urbanisation (item 7) at the higher levels of average national income (Pakistan and Sri Lanka in item 8), but downward at the lower end of national income level.

The relationship of the sex-ratio of the workforce (and female LFPR) with the household economic level and the level of female education shows departures from linearity which, whether or not one likes to describe the curve as a modified U-shaped one, need to be explained. In each of the two years under consideration, the female percentage of the workforce was larger in Nepal and India, both relatively more rural/agricultural and having lower

43

Table 1.1 Inter-country and inter-temporal workforce participation of females and related factors in South Asia

	Year	Bangladesh	India	Nepal	Pakistan	Sri Lanka
1. Female % of workforce (10+ years)	1960	4.6	31.6	35.4	8.0	24.8
	1980	6.3	27.2	34.9	10.4	26.9
2. Female % of agricultural workforce	1960	4.4	35.8	36.6	9.3	29.0
	1980	6.1	30.9	36.5	8.2	29.2
3. Female % of non-agricultural workforce	1960	5.8	19.9	15.7	6.0	19.3
	1980	7.0	18.7	14.8	13.1	24.2
4. Female LFPR (10+ years)*	1960	4.8	40.0	47.6	7.9	27.1
	1980	5.6	29.7	43.6	9.3	26.5
5. Female LFPR (15+ years)	1960	5.0	42.6	49.4	8.3	31.4
	1980	9.8	32.2	45.4	10.6	30.7
6. % of female workers in agriculture	1960	82.3	83.7	97.5	70.8	66.2
	1980	72.0	79.2	97.0	43.0	58.0
7. Urban % of population	1960	5.2	17.9	3.1	22.1	17.9
	1980	10.4	23.4	6.1	28.1	21.1
8. GNP per capita (in $US)	1980	130.0	240.0	140.0	300.0	270.0
9. Female literacy rate % (15+ years)	1961	–	13.2	–	7.4	63.7
	1981	18.0	24.8	9.2	15.2	81.2
10. Gross primary level enrolment girls as % of age group	1960	27.0	40.0	13.0	21.0	90.0
	1982	49.0	70.0	43.0	33.0	101.0
11. Gross middle to secondary level enrolment of girls as % of age group	1965	3.0	13.0	2.0	5.0	35.0
	1985	10.0	24.0	11.0	9.0	67.0
12. No. of small children (male + female up to 9 years) per 100 female population 10+ years	1960	93.0	84.0	75.0	104.0	92.0
	1980	104.0	76.0	94.0	94.0	64.0

Source: United Nations: *World Demography Estimates and Projections, 1950–2025*. A report prepared jointly by the UN, the ILO and the FAO, 1988.
ILO: 'Economically active population: estimates and projections 1950–2025', Table 2, 1986.
UNESCO: Report of the Asian Regional Seminar on access of girls to primary education, 1979.
UNESCO: *Development of Education in Asia and the Pacific: A Statistical Review*, 1985.
The World Bank: *World Development Report*, 1981, 1989.

Note: * The figures here are higher than the activity rates published in the ILO and UN source quoted below because I have divided the economically active females 10+ years by the female population 10+ years, instead of dividing by the total female population (including children under 10 years) as these sources do.

female incidence of literacy and school education than in Sri Lanka. The smaller difference between Nepal and India, which actually increased over time, can partly be explained in similar terms of differential per capita income and female incidence of education. Over the twenty-year period, as the extent of urbanisation and non-agricultural employment options as well as the pool of primary- to middle-educated young women increased, more or less in all the five countries, the female proportion of the workforce decreased in Nepal and India where the female LFPR had initially been higher and mostly in agriculture. The female proportion of the workforce increased where (particularly in Sri Lanka) the primary- and middle-educated proportion of women was and continues to be much higher. It increased also with the growth of non-agricultural employment options which absorbed girls and women seeking to earn but not incur the stigma of field labour. This (as in Bangladesh and Pakistan) explains the much lower initial level of female LFPR that is not entirely due to under-enumeration of female labour.

The second observation that can be made from the patterns of variation in Table 1.1 is that the ratio of small children, i.e., below 10, to older girls and adult women (who take care of them) is very large in Bangladesh and Pakistan in both the years. This is probably one of the important reasons both for the lower economic activity rate of women, except under pressure of extreme poverty forcing childcare to give way, and for the lower rate of school enrolment and completion among girls of 10–14 years. In 1960, the ratio of small children to female population of 10 + years was also quite high in India, Nepal and Sri Lanka. However, by 1980 it had decreased dramatically in Sri Lanka along with a sizeable drop in fertility. In India, this decrease was marginal as was the decline in fertility. On the other hand, in Nepal, the ratio increased with little change in fertility. Social policy thus seems to be more important than income increases in promoting girls' schooling. Pakistan with the highest per capita income of the five countries fared worse than not only Sri Lanka, but also India and Bangladesh, and was nearly at the level of Nepal with half the per capita income level.

The third observation from Table 1.1 is that in the countries (Sri Lanka, Pakistan, Bangladesh) where the female juvenile (10–14 years) rate of work participation has been low, though for different reasons, the gap between LFPR for 10 + years and LFPR for 15 + years (items 4 and 5) is proportionally less than in countries (Nepal and India) where juvenile girls have been and still are more heavily engaged in economic activity.

The fourth observation is that between 1960 and 1980, the percentage of the female non-agricultural workforce increased in Sri Lanka and, to a lesser extent, in Pakistan and Bangladesh, but decreased slightly in Nepal and India. The percentage of the female agricultural workforce either increased slightly or remained unchanged except in India where it fell slightly. Actually, as noted in the following section, the percentage of female agricultural workers in India decreased over the 1960s, mainly with the withdrawal of female family labour from farm work by the newly affluent farmers. However, it increased to some extent over the 1970s mainly with greater male migration from peasant families and/or local switch of male labour from subsistence farming to cash crops or small business, leaving women with more of the work of subsistence production. The increase in female LFPR over the 1970s on this account was tempered by the modest decline in the poverty ratio between 1973–4 and 1983 which slowed the rapid increase of women in agricultural labour experienced in the 1960s without quite returning to the initial level.

INTERSTATE ANALYSIS OF ECONOMIC ACTIVITY RATE AND FEMALE PROPORTION OF WORKERS IN INDIA

Let us now look at the data relating to India, the census data adjusted for comparability over time, and the National Sample Survey (NSS) estimates which also permit some long-term comparison. Over time and across the states of India, we find patterns that parallel, at least are consistent with, the patterns noted in the earlier section from inter-country data.

Between 1960–1 and 1968–9, the incidence of rural poverty increased in India[2] and in many areas also the incidence of operational landlessness, even as the spurt of the 'Green Revolution' started in a few states in the second half of the decade. The proportion of rural households, having to depend on the fluctuating, uncertain, low-wage casual labour market increased. The growth of agricultural employment and the options for seasonal migration of labour were generally too inadequate relative to the influx of the asset poor, especially the female asset poor, into the labour market. The female workforce participation rate (WFPR) decreased over the 1960s in almost all states. At the same time, the proportion of women working in agricultural labour and other low-wage casual jobs increased.

Over the 1970s, say, between 1973–4 and 1983, the incidence of rural poverty decreased for India as a whole.[3] Across the states it decreased the most in Kerala, Andhra Pradesh, Madhya Pradesh, West Bengal, Orissa, Karnataka and Gujarat.[4] Over the long haul, from the poverty estimates for the four normal weather years 1960–1, 1968–9, 1973–4 and 1983, the states that persisted in having the highest poverty ratio are Tamil Nadu, Bihar and Orissa.[5] The high-income, high-growth states of Punjab and Haryana throughout had the lowest incidence of poverty, despite its going up and then down over the decades. Apart from these two states with the most favourable resource–population ratio, the states with the lowest poverty ratio were Gujarat in all the four years, joined by Assam and Rajasthan in 1968–9, and by Kerala, Karnataka and Andhra Pradesh in 1983. Of these, Kerala and Andhra Pradesh are much poorer than the others in terms of per capita state income, showing the feasibility of lowering absolute poverty even at a quite low average state income level. The poverty ratio in West Bengal moved from relatively low in 1960–1 to relatively very high in both 1968–9 and 1973–4 to moderately high in 1983. Tamil Nadu, with its masses of landless labourers, managed to reduce its persistently high poverty ratio only by 1983, and still it is a little above the average poverty ratio for India as a whole.

Between 1971 and 1981, while there was some decrease in the incidence of poverty in India as a whole, more in some states and less in others, there was an increase in the percentage of main workers in the female population in both rural and urban areas.[6] This occurred both in states with traditionally low female WFPR (Haryana and West Bengal) and in those with traditionally high female WFPR (Tamil Nadu and Karnataka) (Banerjee 1989a). The number of main female workers per 100 male workers increased (1961–71) from 23 to 29 for rural India as a whole. Across the states in 1981, this ratio was over one-half in Maharashtra and Andhra Pradesh at one end, and under one-sixth in Punjab, Haryana, Uttar Pradesh and West Bengal, despite the increase since 1971. The female–male ratio of marginal (part-time) workers also increased, especially in rural India. This ratio among main workers was lower in urban India and increased less, except in Kerala.

Part of the increase in female LFPR over the 1970s was due to a rise in female juvenile (under 14 years of age) labour, mainly in subsistence farming (most of all in Punjab, Orissa and Tamil Nadu). Sundaram (1989) has argued this to be the product of three factors: drop in fertility resulting in girls' shift from childcare work to direct earning, increase in productive asset holding, increase in household-based work on land and animal husbandry, and fall in real income of the families at the lower end which depresses schooling of girl children further. Banerjee (1989a) counters these points by noting that since the asset base per family has not increased, at least not in land, while female juvenile participation increased the most in farming, and since low income would not explain the drop in male juvenile participation in cultivation over this period, the change can be seen more as family strategies of shifting boys from cultivation to training or migration for diversifying earning sources, and engaging girls more in subsistence farm work and direct earning beyond helping in mothers' work. A significant part of the rise in female WFPR over the 1970s can be accounted for by the increase in the males from peasant families either migrating for urban jobs or undertaking commercial crops and business as the rural growth environment improved over wider areas towards the end of the decade.

The rise in female juvenile and adult WFPR in the 1970s would thus seem in part to have been a derived effect of family strategies centred on male occupational advancement. In part, the rise for the more affluent might have been due to the growth of both rural and urban semiskilled jobs for middle-educated women in public services and administrative sectors and in the expanding small-scale industries for export and the domestic market.

The important point to remember, however, is that the rise in female worker–population ratio in the 1970s as recorded by the census was not enough to match the level at the beginning of the 1960s.[7] The NSS estimates, on the other hand, show a more moderate rise in the female WFPR over the 1970s. On considering both data sources, it seems reasonable to conclude that the much higher female worker–population ratio in rural as compared to urban India and the one which carries more weight in the national average, increased moderately over the 1970s. However, over the 1960s and the 1970s, it either did not rise at all or decreased a little. This is consistent with the evidence of the rural poverty ratio rising in the 1960s and falling in the 1970s, and the increasing percentage of primary-to-middle educated young women, who would be inclined to shun farm work and low-wage menial jobs and move towards either work in white-collar jobs, in export industries (garments) and modern small industries (electronics), or prepare for upwardly mobile marriage, or do both.

Looking at the pattern of variation across the Indian states in the female WFPR in 1983 in relation to the corresponding patterns of variation in some of the explanatory factors, one can make the following observations from Table 1.2.

First, the female WFPR is consistently higher in rural than in urban areas. However, the states most industrialised and urbanised (Gujarat, Maharashtra and Tamil Nadu, all with more than 30 per cent urban population in 1981) also had in 1983 relatively high rural female WFPR, possibly due to the rural linkages of urban/industrial growth and to the larger proportion of primary-educated women with access to both public- and private-sector jobs requiring some education.

Moreover, in Tamil Nadu and Maharashtra, the rural working women expressed greater willingness to take up additional employment if it was available (as shown in the column on underemployment). The states with the least degree of urbanisation (Bihar, Orissa, Assam and Uttar Pradesh with 18 per cent or lower urban population in 1981) had rural female WFPR much lower than the national average. Further, the women workers did not indicate much readiness to seek additional employment, discouraged probably by the depressed economic environment (relatively low agricultural growth and weak urban-originated growth linkages). These states, also had a lower proportion of primary-educated rural females.

The most outstanding exception to this pattern is Kerala which despite a relatively low urban/industrial development, had by far the highest percentage of primary-educated females, high rural WFPR and a high proportion of the working women expressing dissatisfaction with their underemployment and willing to take up more and better work. The low urbanisation in Kerala is partly due to the state's fairly successful programme of homestead

Table 1.2 Inter-state variation in women's work and some related factors in India

State	% 15+ female population with up to primary education in 1983		Work-participation rate (principal and secondary) Female 5+ yrs 1983		Peasants with <1 ha operational holdings		Lab hhs as % of rural 1977–8*	Underemployment of usual-status workers, 5+ yrs female 1983	
	Rural	Urban	Rural	Urban	% change 1970–1 / 1980–1	% in all holdings 1980–1		Rural	Urban
All India	13.3	24.0	39.3	17.3	+40	56	36.8	18.1	14.6
Andhra Pradesh	10.3	23.4	54.4	20.6	+59	52	46.7	21.5	15.5
Assam	24.5	32.1	14.5	8.5	+22	59	30.1	8.0	4.7
Bihar	5.6	15.0	29.0	12.2	+75	76	41.1	20.4	23.3
Gujarat	18.1	28.7	46.7	15.1	+23	24	36.8	9.9	15.3
Haryana	8.1	17.9	27.5	12.4	+30	32	26.7	7.1	7.3
Himachal Pradesh	20.3	19.9	54.3	19.2	-0.6	55	13.5	7.8	16.6
Karnataka	13.6	22.2	45.0	22.5	+38	35	45.2	20.0	10.7
Kerala	41.1	35.4	35.0	24.3	+55	89	47.4	27.3	20.2
Madhya Pradesh	8.4	22.3	50.9	16.8	+25	33	30.9	9.6	7.0
Maharashtra	16.3	25.0	54.4	17.2	+63	29	46.3	24.0	17.6
Orissa	14.2	22.3	33.4	12.8	+6	47	41.7	21.4	16.5
Punjab	17.5	18.5	36.5	14.9	-62	19	31.8	14.6	9.0
Rajasthan	5.3	16.1	54.6	22.7	+40	29	16.1	7.4	7.7
TamilNadu	18.8	29.7	51.4	23.8	+59	69	47.8	30.1	21.3
Uttar Pradesh	8.5	16.9	30.1	11.2	+20	71	22.9	8.8	8.9
West Bengal	20.1	33.8	22.2	14.5	+62	70	44.7	37.1	15.8

Source: National Sample Survey Organisation, Sarvekshana, April 1988, XI (4), Tables 7, 11.1, 31. National Sample Survey Organisation, Sarvekshana, Jan.–Apr. 1985, VIII (3 and 4).

Note: * Figures of rural households as % of all rural households in 1977–8 are from Table B, p. 6.

allotment for the rural landless which has discouraged their permanent migration to cities, further helped by the state's excellent bus transportation allowing a greater degree of rural–urban commuting for work. Although stressing and accomplishing more in the areas of education, healthcare and poverty-amelioration measures like subsidised food rations, home-site for the rural landless and support for unionisation of farm labourers, and coir and bidi workers – all with large female proportions – Kerala's public policy over time had done much less than some other states to promote rural infrastructure and incentives for agricultural growth. Policies promoting a higher rate of agricultural and industrial growth could in this case both push up further the rural female WFPR and reduce the incidence of poverty.

Another point from Table 1.2 is that, across the states, variations in the rural female WFPR are directly related to variations in the proportion of small-holding peasantry and near-landless labour households. The peasants operating up to one hectare of owned and/or rented land usually have most of the women working on farm operations and household industry. The landless or near-landless labour households account for almost all of the large number of female agricultural and construction labourers and servant maids. States with larger proportions of either of these two partly overlapping categories of rural households would seem more likely to have higher female WFPR, though mostly in low-productivity or low-wage work. Similarly, states with lower proportions of either category would seem more likely to have lower female WFPR. Table 1.2 corroborates this to some extent. Five states with relatively high proportions of either of these two categories of rural households (Tamil Nadu, Andhra Pradesh, Karnataka, Maharashtra and Himachal Pradesh) do have relatively high rural female WFPR. Three states with relatively low proportions of these two categories (Punjab and Haryana have a low percentage of agricultural households and < 1 ha holdings, and Uttar Pradesh has a low percentage of the labourer households, but ranks fairly high in terms of the percentage of < 1 ha holdings) have relatively low WFPR. The major exceptions to this relationship are at one end, Kerala, West Bengal and Bihar, which despite having the highest incidence of small peasantry and/or labour families, did not have a very high rural female WFPR. At the other end, Rajasthan, Madhya Pradesh and Gujarat, which despite their low incidence of small peasant and/or labour families, had rather high WFPR. The first set of exceptions is probably explained by sluggish state growth environment,[8] and by the fact that peasants with nearly one hectare in the first two of the states in question are really small farmers who can afford to spare young females from farm work for schooling and preparing for better jobs or upwardly mobile marriage. Some of the exceptions in the second set can be explained by the lower incomes of their small holders, by the greater amounts of female labour that have to be spent in animal husbandry and in subsistence chores like water and firewood collection, by the relatively low percentage of women with some (up to primary level) education disinclined towards farm-related work, and by a relatively weak growth environment depressing the productivity of self-employment even in larger farms and entailing the greater extent of their engagement in low-yield activities.

The third point is regarding the inter-relationship between the incidence of poverty, the incidence of landlessness, the level of female WFPR and inversely, of female underemployment. This inter-relationship has been the subject of in-depth studies, and I find some of their results broadly corroborated in Table 1.2. First, unlike the primarily self-employed, the labour households with little or no land account for a disproportionately large section of those below the poverty line and an even larger proportion of the total unemployed days for female workers (Tendulkar and Sundaram: 1988). Second, higher proportions of female than male casual labourers are from households living in absolute poverty. Table 1.2 makes it quite clear that the four states – Tamil Nadu, Andhra Pradesh, Bihar and West Bengal – with very high percentages of labour households also have very high incidence of unemployed days reported by even the usually working women, and three of these states have poverty ratios higher than the Indian average. Kerala, despite the very high degree of rural proletarianisation and the associated high rate of unemployment, managed to lower the poverty ratio chiefly with the programme of subsidised food rations for the rural labour families, improving their real wages. At the other end, the states with low percentages of rural labour households – Punjab, Haryana, Assam and Rajasthan – have lower rates of reported female underemployment and lower poverty ratio with the exception of Rajasthan in the latter case.

Regionally, high incidence of smallholding peasantry and lack of economic growth tends to generate higher female workloads, both in the conventional forms noted in census data and in the forms of substance processing and collection often missed by the census. High incidence of wage-dependent landless households generates two kinds of responses to the poverty and seasonal unemployment. One is greater resort to seasonal labour migration which would show up as a higher economic activity rate in census or household survey data. The other response is greater resort to the time-consuming and laborious gathering and scrounging activities, especially in an environment of poor availability of subsistence resources from the commons and forests. Both these responses burden women with back-breaking labour, without making a significant dent in their poverty. High incidence of middle-to-large farmers in a region, as in Punjab, Haryana and the Pakistani Punjab, with low incidence of landlessness tend to produce a low female worker rate, especially in farming unless there is enough expansion of at least primary education among the daughters and wives of affluent peasant families and of rural job opportunities for them.

Table 1.3 Education-specific WFPR for female population 15+ years in India, 1983

	Illiterate	Up to primary	Middle	Secondary	Graduate and above	All 15+
Usual workers						
Rural female	40.7	27.6	15.8	15.8	22.0	37.3
Urban female	24.1	13.4	7.1	13.4	27.9	18.0
Rural male	91.0	88.5	70.2	66.5	77.4	85.6
Urban male	86.3	85.1	63.8	65.2	81.0	76.9
Usual and subsidiary workers						
Rural female	54.3	41.8	29.3	26.9	29.5	50.9
Urban female	29.6	18.1	11.1	15.6	29.7	22.5
Rural male	91.8	90.0	75.3	73.3	83.9	87.7
Urban male	86.8	85.2	71.0	67.5	82.5	78.4

Source: National Sample Survey, *Sarvekshana*, April 1988, XI (4), Table 12. NSS 38th Round, 1983.

EDUCATION AND ECONOMIC ACTIVITY RATE

In the section on the inter-country pattern of variation (pp. 41–6), a non-linear (sort of U-shaped) relationship between female WFPR and the incidence of literacy and education levels is noted. In Table 1.2 it is again noticed that the states where the percentage of females with primary education is considerably higher than the national average do not rank very high in rural female WFPR. On the other hand, some states with a very low percentage of such females have high WFPR. In Table 1.3, this issue is further explored with the help of NSS data for 1983 on education-specific female WFPR.

Clearly, a much higher proportion of illiterate women work, presumably for very low and irregular insecure wages and low-productivity categories of self-employment. In rural areas, with more opportunity for self- or family-employment in land and in household industry, and with more demand for part-time female labour in farm-related operations, the work-participation rate is the highest among the illiterate. With some education, which is correlated with better income level, there is a sharp withdrawal of female labour from low-wage unskilled labour and also from some kinds of drudge work for which servants or private facility (for water etc.) are used while purchasing ready-made foodstuff substitutes home processing. The women with middle-to-secondary education shun the kind of work that the illiterate poorer women or even their own illiterate mothers may be doing. If suitable white-collar or semiskilled job opportunities are not available, they prefer and can more often afford to be full-time housewives. Being spared the aggravating load of low-return work that the illiterates must do, carries status value and benefits in terms of leisure, time for family care and help to small children in their education.[9] In the case of males, although the economic necessity to work for extremely low wages or at laborious but low-yield tasks diminishes for those who have the primary and middle education, the drop in WFPR is much less pronounced because of greater access to more job options locally or as migrants. At higher education levels, WFPR rises for both males and females in professional and regular service jobs in rural and urban areas. But for females this rise is less, because they are more constrained in their access to such jobs and by the claims of reproductive labour which is partly but not entirely eased by servants and the help available within the extended family.

The inter-state variation in the pattern of education-specific female WFPR shows some significant deviations from the overall pattern mentioned above.[10] The drop in female WFPR with the attainment of primary education is very little in Kerala, no more than for the male. This may be because in this state, despite low income levels, the overall female literacy and education up to primary level are not only the highest in India, but they also no longer signify the work-status hierarchy.[11] The WFPR of the illiterate urban females in Kerala, as also in the three other south Indian states in 1983, is much greater than in the rest of India, especially Uttar Pradesh and Haryana. The WFPR of illiterate rural females in Kerala is less than in India as a whole, much less than in Madhya Pradesh, Rajasthan, Maharashtra and also the other southern states (where the spread of rural industrialisation has been greater, whereas in Kerala older rural industries like coir processing are declining and fewer new ones have arisen due to the past neglect of developing infrastructure and encouraging the expansion of household industry and industrial subcontracting, where the other southern states have gone much further ahead).

INTER-COUNTRY AND TEMPORAL PATTERN OF FEMALE LFPR BY AGE-GROUP

A set of graphs in Figure 1.1 shows the economic activity rate of females by age group in five countries in 1960 and 1980. It reveals some inter-country and inter-temporal differences which are significant.

First, the economic activity rate of juvenile females (10–14 years) is much less in Sri Lanka due to the much greater school enrolment and completion rate of girls, and in Pakistan and Bangladesh, due not just to lower marriage age which is the case also in Nepal and India without producing a similar low economic activity rate of girls, but also to the stronger culture of purdah for pubescent girls. In India, nearly a quarter of girls aged 10–14 worked in 1960. In 1980, possibly due to the moderate increases in household income and in the school enrolment of girls, this proportion decreased to one-sixth. In Nepal, over a third of the girls aged 10–14 worked in 1960 as well as in 1980 which is consistent with the much greater share of agriculture in the workforce, lesser agricultural development, lower levels of school enrolment, the relative absence of taboo against women farming and marketing and last but not the least important, the much greater amount of collection work for firewood, fodder and water done largely by the juvenile females. The last factor also explains part of the rather high WFPR of girls in India.

One may argue that the incidence of female child and juvenile work is lowered by increase in family income, by increase in the schooling of girls, which is only partly linked with increase in income, by higher female age at marriage and also by the nature of patriarchal restrictions, whether Islamic or Hindu upper caste. Within India, the economic activity rate among girls aged 5–14 is much lower than the national average in Kerala and is comparable to Sri Lanka because of the higher rate of mothers' literacy, the sustained state policy and social movement supporting girls' schooling.

The second point highlighted by the graphs in Figure 1.1 is about the differences in the range of ages for the peak of economic activities by females. In Sri Lanka, with greater incidence of girls' schooling and later marriage, the peak was reached in the age group 20–24 in both 1960 and 1980, and it was more or less maintained through to age group 50–54, after which it declined, more sharply in 1980 than in 1960, possibly reflecting a slump in the export and slack labour market in the later year. In India, the top range of the rather smooth curve for economic activities by females spanned 25–49 years in 1960, but in 1980 the top range was wider starting from 25 years of age. In Nepal, in both 1960 and 1980, the peak for economic activities was reached in age group 15–19, after which the rate decreased somewhat in the prime child-bearing age group ranging from 20–29, years remaining steady until age 59, and then it slowly came down back to the juvenile rate of about a third. Evidently, old women and young girls work a lot more in Nepal than in India or Sri Lanka. In Bangladesh, the peak activity rate was reached for women in their thirties, past the prime child-bearing age, after which it started to taper off. There is not much change in this pattern between 1960 and 1980 except that in 1980 there was a pronounced spurt in the 20–24 age group. This was most probably due to the expansion of light export industries hiring young women for a few years before they would be married off. The same factor would explain a similar spurt in worker ratio in the 20–24 age group in Pakistan where the peak range of economic activity occurred in the age group 25–59 years in 1960, and expanded a little in 1980 to 20–59 years.

EMPLOYMENT PATTERN AND OCCUPATIONAL DISTRIBUTION OF WOMEN WORKERS

The distribution of women's work by the sector of production (agriculture, industry, trade and services) and by the mode of employment (self-employment, work in family enterprise and wage-employment further divided into casual or regular employment) brings out certain patterns across the South Asian regions, the Indian sub-regions and in the changes over time. These patterns reflect the varied working of mainly three factors: a) the regional levels of agricultural development and common property resources for subsistence, b) the nature of male migration from a rural region (pull-type or push-type), and c) the changing configuration of the industries (which hire females generated by the mix of declining traditional and growing non-traditional manufacturing sectors. The modernisation of non-agricultural occupations to the extent that it involves some level of education or training, is associated either with the inter-regional differences in the proportion of females with middle-level education or with the degree of migration of female labour with middle-level education from one region to another, a relationship noted earlier.

Let us first compare the distribution of female workers by sectors of production within India. From items 2, 3 and 6 in Table 1.1, it is clear that in 1980, after three decades of industrialisation, nearly eight out of every ten female workers were still in agriculture, engaged either on the family farm or as hired agricultural labourers or both. The decrease in this proportion has been very slight since 1960. In Nepal, the proportion of female workers in agriculture had remained even higher at 97 per cent. In both Sri Lanka and Pakistan, this proportion was lower in 1960, more so in 1980 (58 per cent and 43 per cent respectively). The reasons are partly different: stagnation of tea export and growth of other industries essentially employing women (in Sri Lanka), the Islamic taboo against women working outside home and the adaptation of manufacturing modes to tap female labour by arranging segregated work spaces. In Bangladesh, as in Nepal, the high proportion of women workers in agriculture is primarily due to the huge share of agriculture in the economy, the lack of growth in agriculture and the prevalence of traditional labour-intensive technology in most agricultural operations, especially those carried out by women. However, unlike in Nepal, the proportion decreased significantly by 1980 in Bangladesh. This is possibly due to the cultural factor of high income elasticity of withdrawal of women from farm work and the recent growth of high export industries hiring the growing pool of middle-educated daughters.

The female percentage of the agricultural workforce has remained around one-third of the total in India, Nepal and Sri Lanka, whereas in Bangladesh and Pakistan the percentages are under ten. Notwithstanding the under-estimation of female participation in farm-related work done within the

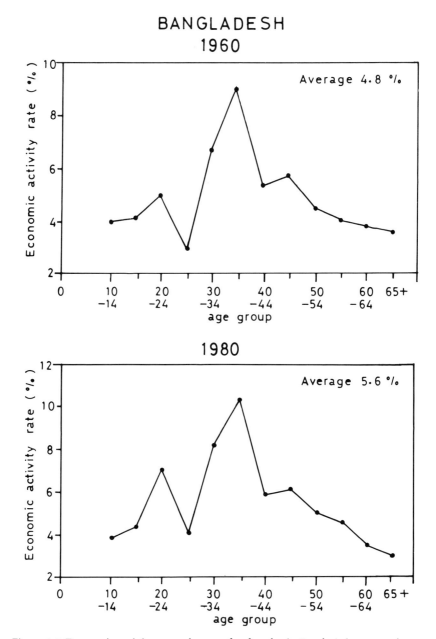

Figure 1.1 Economic activity rates, by age, for females in South Asian countries

Source: ILO: *Economically Active Population, Estimates and Projections, 1950–2025,* Table 2, Geneva: ILO (1986).

Note: The averages in the graphs denote overall economic rates for females 10+ years.

INDIA
1960

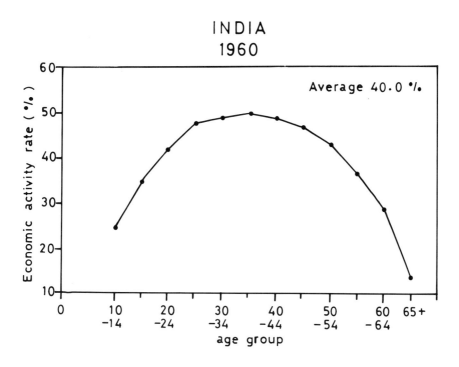

1980

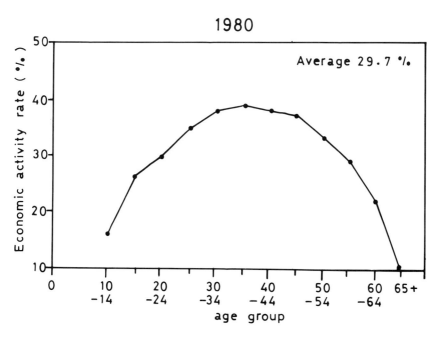

NEPAL
1960

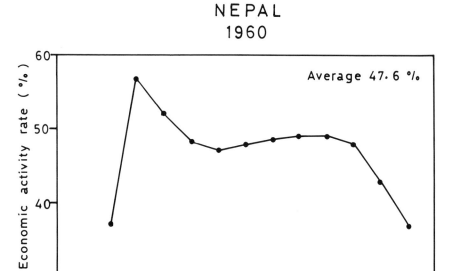

1980

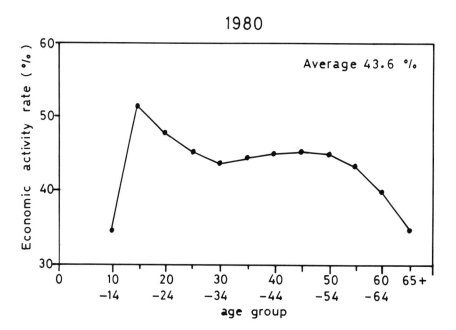

Figure 1.1 continued

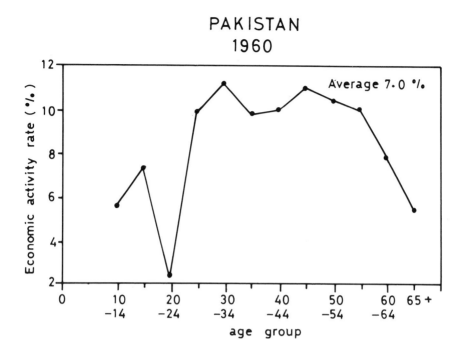

PAKISTAN
1960

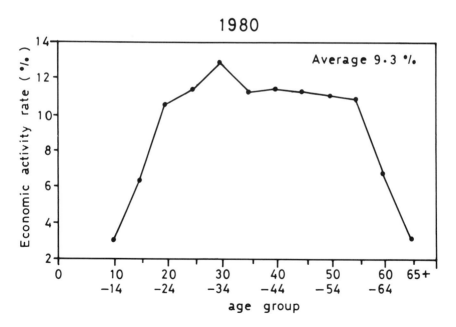

1980

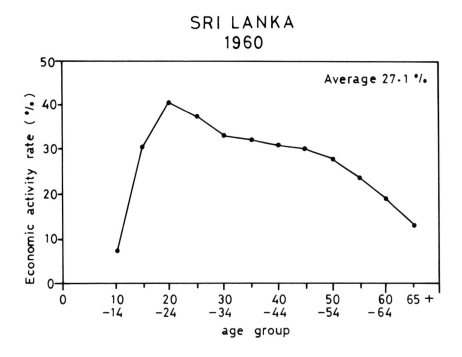

SRI LANKA
1960

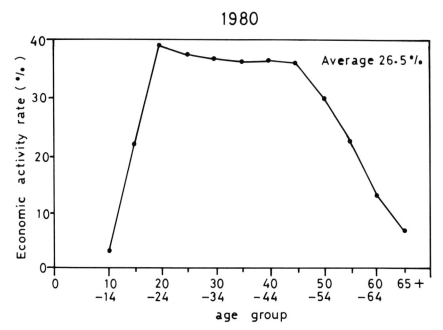

1980

Figure 1.1 continued

home compound, it seems doubtful if Bangladesh and Pakistan would ever approximate the high levels observed in Nepal and south India.

The female proportion of the non-agricultural workforce has been higher in Sri Lanka and India due to a large proportion of female employment on plantations, the vast expansion of the state services and administrative sectors which employ women with various levels of education, and the enormous informal sectors of services and small-scale production units including small-scale animal husbandry where women work, especially in India. The proportion is relatively high also in Nepal, even without the industrial and plantation sectors, due to the greater importance of women in household industry and marketing/trading activities.

Across states within India, the percentage of female workers in agriculture (and the female–male ratio of the agricultural workers) in 1981 was well above the national average in Andhra Pradesh, Maharashtra, Tamil Nadu and Madhya Pradesh. It was well below the national average not only in the high-growth, traditionally male-dominated wheat farming of Punjab and Haryana, but also in West Bengal with a moderate growth in agriculture, perhaps because the stigma attached to women from respectable families doing farm work is nearly as strong as in Bangladesh.

Between 1961 and 1981, the sex ratio (female–male) in agricultural work went up considerably in India: while the percentage of male workers in agriculture decreased, the percentage of female workers therein increased – which over the 1960s came about largely with a sharp increase in female agricultural labour due to increased rural landlessness. On the other hand, in the 1970s there was a rise in female work in subsistence farming due to a spurt in migration and occupational diversification of adult and juvenile peasant males.

Over the 1960s the percentage of women working as agricultural labourers went up the most whereas over the 1970s it was the percentage working in subsistence production. Neither the same factors nor the same category of households were involved in these shifts. Earlier, it was the proletarianising push-pull effects of agricultural growth; the major correlate was the rising incidence of poverty and lack of assets for self-employment. Later on, it was the smallholders' strategies (and greater opportunities) for diversified male earnings. The opportunities arose from the linkages of agricultural growth in some areas and from outmigration in others. The major correlates were some increase in income (or at least in opportunities for non-agricultural earning) for some sectors of smallholders and the pull of a tightening rural labour market. Increase in commercial opportunities for India's peasant men over the 1970s has affected the sectoral distribution of women's work indirectly, by making them take the work left by men and boys (Banerjee 1989a). Rural non-agriculture and medium-term job migration have remained male-dominated, despite the regional prominence of women in animal husbandry, wage-labour in public works and in contract migration.

A set of ILO-sponsored analyses of limited work options of rural Indian women presents an explanation in terms of not only the pattern of economic growth but also the inter-related factors of low female literacy, marketable skills and high fertility rate. It is noted that since a largely illiterate female population with a high fertility rate tends to be confined to the agricultural sector and to casual labour, in order to be effective in the longer run, policies to improve female employment must concentrate on female literacy and skill training, directed especially at girls and younger women, and on fertility reduction (Jose 1989). Although important, these two factors do not quite explain why, for instance, the percentage of female workers in agriculture from 1972–3 to 1983 increased more in Punjab and Haryana on one end, and in Kerala, Bihar, West Bengal and Uttar Pradesh on the other end of the scale of economic growth. None of these states experienced any worsening of female education and fertility levels that might explain this. The factors that are more consistent with this change are a) the increase in rural male migration and shift to non-agricultural occupations, leaving women to work more in subsistence agriculture, and b) the high incidence of poverty among the rural land-poor that forces their women into low-wage agricultural and construction work and domestic services.

The changes over the 1960s and the 1970s in the composition of women agricultural workers between cultivation of family-operated farms and wage-labour show interesting variations across the states in India, variations which correspond with inequality of land and wealth in general. In India the ratio between women working on their family farms and on others' farms for wages decreased between 1960 and 1970 followed by an increase between 1970 and 1980. But this ratio in 1980 was still much lower than the ratio in 1960 (Table 1.4). The sharp 'proletarianisation' of female agricultural workers has been a trend over the last two decades although it slowed down in the 1970s. In general, the reasons for this trend are: the high and rising incidence of poverty in self-employment assets and the persistence of workers' educational disadvantage in the market, and the increasing prosperity of cultivating farmers, spreading, however slowly, over the two decades both across regions and to smaller farmers. A number of states parallel this average picture. However, there are some very striking differences. In the four states in the south, the percentage of cultivators, initially lower than the national average, decreased more steadily. The initially higher percentage of agricultural labour continued to increase more steadily. This was also true of Bihar, Orissa and Maharashtra. In Himachal Pradesh and Rajasthan, in contrast, with a low proportion of the landless and less agricultural growth, women working in agriculture have always been mostly employed in cultivating peasant households and the percentage working as agricultural laborers was very small in both years. In the three states with the highest

Table 1.4 Inter-state variation in women workers in agricultural India, 1961, 1971, 1981

	1961		1971		1981	
	Cultivators	Agricultural labourers	Cultivators	Agricultural labourers	Cultivators	Agricultural labourers
All India	58.9	27.7	32.6	54.4	40.8	47.7
Andhra Pradesh	40.9	42.5	20.8	66.2	25.9	62.0
Bihar	55.8	32.3	18.0	75.9	31.6	60.8
Gujarat	64.9	24.6	38.8	54.0	39.3	48.9
Haryana	82.2	7.8	45.7	53.0	67.3	24.8
Himachal Pradesh	93.3	1.3	90.7	4.1	92.3	1.7
Karnataka	59.9	28.0	26.7	54.2	32.0	53.6
Kerala	17.4	36.5	50.0	53.5	8.9	46.5
Madhya Pradesh	69.7	23.2	43.3	50.6	52.1	40.7
Maharashtra	59.2	35.6	38.0	56.1	45.2	48.5
Orissa	51.3	29.0	21.3	5.7	26.2	57.2
Punjab	54.4	8.1	11.2	21.7	19.2	65.9
Rajasthan	85.2	6.1	68.2	21.9	77.3	15.3
Tamil Nadu	47.3	35.6	22.3	62.2	26.5	60.1
Uttar Pradesh	66.9	21.7	45.2	46.9	54.8	36.6
West Bengal	41.6	26.0	14.9	54.5	20.6	48.2

Source: Indian Census, 1961, 1971, 1981.

Note: The percentages for cultivators and agricultural labourers do not add up to one hundred, because there is a relatively small category of other agricultural workers, including those in animal husbandry, fruit and vegetable growing.

rates of agricultural growth over the two decades (Punjab, Haryana and Uttar Pradesh), the ratio of female agricultural labourers to female cultivators was very low in 1960, but this ratio increased over the 1960s. In Punjab, the increase which continued through the 1970s was so sharp that by 1980 it was 85 : 15, a ratio matched only by Kerala at the lower end of rural economic growth.

Women's share of total manufacturing employment has increased over the 1970s, in both rural and urban areas. But most of this increase has occurred in household industries. Inter-state, women's participation rate in non-household industries went up substantially in Karnataka, Tamil Nadu, Madhya Pradesh and Orissa. However, it did not rise in Gujarat, Maharashtra and West Bengal, states with traditionally high levels of female employment in organised non-household manufacturing. This depressed the rise at the national level. The declining female share in manufacturing employment at the national level was arrested in the 1970s: women lost ground in the textiles mills (cotton and jute), but gained in the new textile product industries (garments and embroidery), electrical appliances, power-loom production of silk and wool fibres, largely organised as home-based or

workshop-based contract work. Women also lost in traditional food processing (grain milling, coffee curing, cashew processing), but gained in new
ones (confection and fish processing), though the overall growth and their
weight in the total in both categories has been low.

As far as employees in the services (that is, leaving out the largely self-
or family-employed) are concerned, two categories have grown the most,
affecting two socioeconomically different sets of women. One is casual
wage-labour in domestic services in urban areas, which increased by
40 per cent between the two census years. The other is regular salaried
employment in services (including public-financed social, educational,
health and administrative services), all involving some education and
training, as the pool of women with middle-to-higher education increased
by 80 per cent in India as a whole. The female to male ratio of workers
in domestic services, rural plus urban, went up from .60 to .88 over the
1970s, the ratio in all government services went up from .29 to .35. In 1981,
the number of women working in domestic and other personal services,
including prostitution, at the bottom end of the services category was 0.83
million and the number in education, scientific and research services at the
upper end was 1.12 million. Across the socioeconomic strata, there is very
little overlap between these two service categories because the mobility
from one to the other is severely blocked by the gaps in levels of education
and marketable skills, which, though remediable in the long run, are evidently slow to change, especially in the case of females.

Let us now consider the distribution of women workers by the mode of
employment. In all the South Asian countries, the division between the
casual and the regular modes of labour hired and between the formal and
the informal sectors of paid employment respectively, has increasingly
been stratified by sex. Women and girls, whether main or part-time
workers, are found to be proportionally more in casual labour than men
and boys, in both rural and urban areas. In the Indian NSS data of 1983, of
all the principal female workers, 42 per cent in rural and 31 per cent in
urban India were casual employees, as opposed to self-employed and regular employees. In the case of casual male workers, the percentages were
lower at 30 and 15 respectively in rural and urban India. In 1983, in three
southern states, i.e. Andhra Pradesh, Tamil Nadu, Karnataka, and in the
western states of Gujarat and Maharashtra, not only was the percentage of
casual agricultural labourers among rural female workers the highest in
India, but it was also much higher than the corresponding proportion
among rural male workers. These states have had some of the highest
incidences of rural landlessness and also fairly high rates of agricultural
growth in recent years. Although the push of landlessness and the pull of
the agricultural growth explain some of the spurt in casual labour in the
rural workforce in these cases, they do not quite explain such a large sex
differential in agrarian 'proletarianisation'. To explain that, one has to bring

64

in, among other things, the differential access to the non-agricultural employment opportunities locally and as migrants.

Among the provinces in Pakistan, Shah (1986) has noted some striking variations in the occupational distribution of women workers, and in the related factor of the educational incidence. As far as the sectoral distribution of working women is concerned, according to the 1975 Pakistan Fertility Survey, about 38 per cent of currently married rural working women of 15–49 years were employed by the family in agriculture, 5 per cent were employed as farm labourers and 57 per cent were engaged in non-agricultural work, within or outside the home. The proportion in agriculture was the highest for Sindi women followed by women in Punjab and North West Frontier Provinces (NWFP). The lowest was for women in Baluchistan. Baluch women have been employed more in non-agricultural production than in services compared with the other provinces. Undoubtedly, the purdah system constrains female work participation outside the home. Based on a multivariate analysis, taking into account the socioeconomic level and the demographic stage of the family, the correlation is still high between purdah observance and low female WFPR: in rural areas 29 per cent of working women and 54 per cent of non-working women observe purdah (Shah 1986: 293–4). However, Shah rightly notes that purdah, although a serious constraint, has also been instrumental in creating demand for female teachers and doctors. They may be a small proportion of the total female workforce, but explain the very high WFPR of the educated women, which is true of all the South Asian countries. As for the provincial variation in the female literacy rate, according to the 1973 HED survey, literacy for rural females 10 years and above was much greater in Punjab than in the other three provinces. Both Sindi and Punjabi urban women had much higher literacy rates than the Baluch and NWFP urban women. In the 1973 HED survey, school enrolment ratio for girls aged 5–9 years in NWFP and Baluchistan was only half to a third of the rate in Punjab and Sind, even though these higher rates are much lower than in Sri Lanka, and even Nepal and Bangladesh. As in the rest of South Asia, more of the female enrollees drop out before grade V. According to the 1976 statistics collected by the Bureau of Educational Planning and Management, the percentage of girls attending grade V was 20 in Punjab, 10 in Sind, and 6 or less in NWFP and Baluchistan. These figures from the 1970s would be determining the provincial differences in the incidence of middle-educated working women and hence the occupational distribution of females today. Although the incidence and the increase in girls' schooling and higher education in Pakistan has been much less than in Sri Lanka or even India, it has still generated a substantial increase in the percentage of girls' schools and the percentage of female teachers at school level in particular, owing to the stronger culture of sex segregation. The increase in opportunities for better-paid teaching jobs, coupled with the positive status of higher education for women, account for relatively high WFPR at a higher level of education which is generally true of all the South Asian countries. Illiterate women, especially those with illiterate and hence low-income husbands tend to have very high WFPR in all the countries and all the Indian states, but their heavy work gets such low returns and is under such degrading and punishing conditions that they quit when the economic level improves.

The significant importance of government in regular salaried employment in the formal sectors has been noted (Edgren 1987), especially in the less developed, largely agricultural countries. The public-financed sectors (public services including education, health, broadcasting, administration and production sectors) have very high shares in employment in these, both for males and females with secondary and higher education. The public sector has a strategic importance for women's access to the modern-sector jobs and to professional positions (the share of women in public services rose from 8 to 11 per cent in India between 1970 and 1982: half of them as teachers and nurses, a quarter as unskilled labourers and a sixteenth as clerical workers (Edgren 1987: 15). In India, total employment in the public sector grew from 7.1 million in 1961 to 17.3 million in 1985 accounting for about 11.7 per cent of total employment in 1981. Although a certain geographical pattern is noted across states (Edgren 1987: 71), the temporal growth in public-sector employment has been the highest in states with higher economic growth, i.e. Punjab, Gujarat, Haryana and Karnataka with the exception of Delhi and the border states. Thus, generally speaking, the public sector did not quite correct the regional imbalances in private industry which tended to go to the more rapidly growing states. On the whole, the public sector employs a larger proportion of white-collar than blue-collar and unskilled workers and this difference has been accentuated over time. In the white-collar, professional and clerical jobs, the share of female employment has fared much better in the public than in the organised private sectors.

In Bangladesh since 1982, as the Martial Law regime attempted to roll back the public sector and promote private enterprise, total employment in the public sector increased from 0.9 million in 1972–3 to only 1.3 million in 1983–4 (Murshid and Sobhan in Edgren 1987: 36–7). Seasonal public employment (in construction, Food for Work and agricultural projects) increased marginally from 1.3 million in 1972 to only 1.75 in 1983–4. Most of this rapid increase was in managerial and administrative categories followed by the category of clerical, professional and technical, and production and transport workers, in that order, the last category having initially been the largest which declined sharply in recent years.

Apart from the direct employment in the public-financed sectors and the question of women's share in such employment, the government has played expanding roles in affecting women's self-employment, its quantity and quality through various programmes of providing loans, and distributing know-how and skills (via sewing machine), installation of accessible sources of potable water and improved cooking stoves. However, the policies that curtail the female poors' access to subsistence resources from the commons act negatively. These programmes and policies have affected women's work differentially in different regions and in different socioeconomic strata. State support or opposition have also helped or repressed the efforts by the female working poor and by social activists to mobilise for the collective amelioration of work conditions, for demanding and securing essential public services, and for curbing the societal and familial sources of their oppression as workers and as women.

THE FEMALE WORKING POOR ORGANISING COLLECTIVELY FOR IMPROVEMENT

In parts of South Asia the last two decades have seen the emergence of a variety of forms in which the female working poor have mobilised for collective organisation to address specific economic, work-related needs and/or broader problems that adversely affect their work conditions and their lives. These various forms of mobilisation have arisen either on their own or with the initiative of visionary individual activists, dedicated NGOs, trade-union activists, political party cadre, and in some cases government agencies.

Ela Bhatt organised thousands of self-employed women garment workers, handcart pullers, vegetable vendors, junk-smiths, milk producers and rag pickers of Ahmedabad, India into the Self-employed Women's Association (SEWA) started in 1972. This was with the initial support of and inclusion in the Textile Labour Association showing that a union could exist for the development of its self-employed members as well as for resisting exploitation of employees (Esterline 1987). Later on, SEWA became an autonomous organisation, it vastly expanded its membership and spread to other cities. It formed a women's cooperative bank, pooling the small deposits of its members, which is recognised by the Reserve Bank. Besides supplying low-interest loans to its members, SEWA also provides guidance in financial management, marketing and purchasing of materials and has, with some success, lobbied and agitated to end police harassment of street vendors, supported sponsored literacy and training programmes and established childcare centres, and negotiated with the state housing board for low-cost housing. SEWA workers assist members in obtaining prenatal care and immunisation of babies.

> Bhatt's optimism is based on the belief that if SEWA, the union, creates the motivation and SEWA, the bank, provides financial and managerial skills, the self-employed worker will soon be able to raise both her economic position and her self-respect. . . . In her 1983

address to the 13th World Congress of the International Confederation of Free Trade Unions in Oslo, Bhatt emphasised that in India, as in many developing countries, [much] of the work force are outside of factories, firms, and farms . . . and the labour movement would remain incomplete if they are not unionised. The Congress passed a resolution . . . noting that such organisation [of the self-employed] was not a threat to organised labour, but complementary to it.

(Esterline 1987: 82, 87)

The Working Women's Forum, started in Madras city and spread in many areas of south India, has its success based on securing loans from nationalised banks for its asset-poor self-employed women on the basis of neighbourhood-based group collateral. Notable among other organisations focusing on the needs of poorer working women is the women's wing of Shramik Sanghatana in Maharashtra, which organised adult night schools and mobilised agitation against alcoholism and gambling, which in poor families drain male earnings, exacerbate women's workload and undermine subsistence. Also notable are the construction labourers' unions in Maharashtra, which have tried to implement the legal reduction of the wage gap by sex and seek childcare for women labourers with small children. The Jharkhand mukti Morcha, originally a militant trade union of peasants fighting for the alienation of their land, has started collective measures for economic improvements like eliminating middlemen, and campaigning against alcohol and costly marriage ceremonies. The Chipko and Appiko movements, basically by women directed against commercial logging and quarrying, but actually addressing rural women's aggravated problems of collecting firewood and fodder caused by deforestation, are well known. The Kerala unions of agricultural labourers, coir processors in small non-factory units, *bidi* rollers and cashew processors, all having large female proportions, have played important roles not only in reducing the wage gap by sex and harassment of women labourers at work, but also in securing government programmes of subsidised food rations and, in the case of landless rural labourers, allotment of home-sites. The union of maidservants in a few cities try to resist arbitrary sacking, wage deduction and withholding. The dairy cooperatives, organised with the agency of the National Dairy Development Board, in west, north and south India with over 2.5 million members, have helped small milk producers, many of them women, by eliminating the exploitative trader-middlemen and by offering a better price to the producer.

In Bangladesh, the staff members of the Women's Wing of the Integrated Rural Development Programme have formed cooperatives of women engaged in subsistence activities to provide credit facilities to develop enterprises in which they already have experience. The cooperatives, although concentrated on supporting individual women's economic projects, have become involved in broader activities, such as functional literacy classes, nutrition instructions and sending young members to train in rural health centres. The Bangladesh Rural Advancement Committee (BRAC), a non-governmental institution started in 1971 for relief and rehabilitation after the 1971 war, has grown into a multifaceted development institution for the poorer rural women, covering more than a thousand villages by the mid-1980s. Moving from the initial extension service approach to a participatory cooperative approach, BRAC has mobilised the poorest groups of women in villages for designing and implementing projects of their own. Inputs and technical services are directed to and managed by the groups with the help of BRAC field staff. For each new scheme, BRAC seeks the help of specialised agencies for training and know-how to build its own technical capacity if the scheme promises to be viable. The cooperatives pool small savings as a loan fund for members to tide them over hard times. What BARC recognises as remaining to be done is finding new employment opportunities for poorer women through integration into the growth programmes, and finding ways to improve the efficiency of women's subsistence work to lessen the time and strain of routine chores (Chen 1983). Like WWF, BRAC also arranges bank loans for members on group collateral, but unlike WWF, BRAC stresses the production development side or the utilisation of the credit. The cooperative project of the Bangladesh Academy of Rural Development (BARD) attempted a large network of loan funds or credit cooperatives formed out of small savings by poor women. However, the BARD credit network was not concretely linked up with a production programme of enterprise development of the kind that BRAC has attempted by arranging concrete help in women's enterprise or in existing self-employment the SEWA way.

This description, illustrative rather than exhaustive, of recent developments in grass-root organisation of the female working poor indicates an emerging institutional source of positive amelioration of the female condition of huge quantities of work at very low return or wage. When these organisations meet with the needed supply of inputs and support (government, NGO and individual activist support) the result can be very encouraging. It may not make poor women rich, but it can improve their productivity and/or wages in their existing work, their self-esteem, ability to resist exploitation and demoralisation in the kind of work they have to do, and generate a collective voice for the provision of the needed basic facilities and access to the existing financial and technological infrastructures established by the government. The cases of such organisation, those that do seem to work, are still very few and regionally confined, but they are potentially replicable.

CONCLUSION

In this chapter, I have examined, with South Asian inter-regional and inter-temporal data, first, the ways in which rates of economic activities by females (LFPR and WFPR) are related to four economic-structural-demographic factors and one religious-cultural factor. The set of four factors are: a) the household income and asset level (land and non-land material assets); b) the level of female education and marketable skills; c) the woman's age, and overall demographic features like the ratio of small children to female juveniles and adults; and d) the environment of economic growth or stagnation in which the women in a particular socioeconomic stratum are located, a composite factor comprising the agricultural growth rate, the options for rural non-agricultural employment, the extent of nearby urban-industrial development either generating growth linkages spreading villageward or pulling rural migrants. The main religious or caste-based cultural difference affecting female LFPR is related to the stigma attached to women visibly doing agricultural work outside home. This factor has been relatively strong in Bangladesh, Pakistan, and north and east India, and relatively weak in Nepal and south and east India.

In identifying the variations and explaining the relationships, I have also tried to make sense of and to some extent reconcile, some of the seemingly opposing directions in which the findings of many of the studies on women in South Asia sometimes seem to point.

One of the conclusions of this chapter is that the asset-poor women with the lowest education level tend to have the highest WFPR/LFPR. A distinction may be made here between households which are smallholdings and landless or near-landless households. Apart from many operations directly or indirectly related to cultivation, the women among the former category clearly have to do a great deal of home processing for family use and household manufacturing; women among the latter group have to depend on fluctuating, casual labour at low wages. They also have high WFPR and enormous time and effort is spent chasing insecure daily-wage work where there is a high level of seasonal unemployment. Both groups struggle with crushing burdens of work which rarely gets them out of poverty, because of the low returns or low wages in such work. If they are located in areas of agricultural stagnation with no nearby locus of industrial/urban growth and if their forest-based subsistence resources are shrinking, then their workloads can become truly crushing, for adults as well as children. Unquestionably, the very high WFPR of the asset-poor, uneducated women in economically deprived or environmentally degraded regions is a serious indicator of underdevelopment. In the country, the sub-region or the economic stratum in which the female WFPR is high largely due to these factors, one would expect economic and educational improvement to bring down the specific WFPR. Being able to withdraw from excessive labour under punishing conditions is an improvement, personally and socially.

The second conclusion is that since the scope for household-based employment is greater in peasant agriculture, a country or region with a relatively high agricultural share of the workforce and also a high incidence of rural landlessness tends to have relatively high female WFPR. Conversely, a country or region with a high rate of urban/industrial expansion, relatively affluent farmers and a low incidence of landlessness tends to have relatively low female LFPR.

The third conclusion is that the income elasticity of (propensity with income rise for) withdrawing family women from farm work and menial labour is closely linked with the female education level, the availability of cheap female labour for hire (which is determined by the extent of asset poverty and the growth or stagnation of employment opportunities for the asset poor) and the religious-cultural value of affording such withdrawal.

The fourth conclusion is that the relationship between female WFPR and the two related factors of family income and level of female education is a non-linear, modified U-shaped one. In a country, region or social stratum in which the income has risen from a very low level and the female education level has moved from illiteracy to primary education, the WFPR tends to be lower. But as family income increases further, and as the female education level rises beyond the primary level, and if job sectors, public or private, hiring middle-to-higher educated women expand, the female WFPR (and LFPR) increases quite consistently in almost all the data.

The fifth conclusion is that the age distribution of females in economic activities shows distinct, and in some ways disturbing, patterns. The WFPR for the juvenile and the old are much higher in India and Nepal than in Sri Lanka, with by far the highest rate of girls' school enrolment and completion rate. High juvenile and old female WFPR obviously have the adverse effects on girls' schooling, women's socialising and childcare roles, and the minimum need for rest. Poverty, private at the household level and public at the environmental and infrastructural levels, increases the load of routine subsistence chores, from collection of water and firewood to home processing and crude methods of cooking and cleaning, a large part of which has to be borne by young girls and old women, especially where the WFPR is very high for women aged 15–54 years and the ratio of children under 10 to adult females is high too. This is the case in Nepal and parts of India, such as Himachal Pradesh, Andhra Pradesh and Maharashtra. The economic activity rates of juveniles and old females in Pakistan and Bangladesh are much lower, partly because of cultural constraints and partly because of relative stagnation of employment options and the low female educational level. Recent expansion of light export processing industries hiring young women with middle-level education for just a few years has increased the LFPR in the 20–4 age group in both countries.

The sixth conclusion is that the casual informal sectors with low-wage and insecure labour, in both agricultural and non-agricultural activities including services, have a higher and rising female to male ratio compared to the sectors with regular employees or self-employed personnel. This is generally true in all the South Asian countries and across the Indian states. This has been happening in areas of economic growth and in areas of stagnation, but with an important difference. The difference is that the informal and casual-labour sectors generated through the linkages of agricultural growth and/or urban/industrial growth offer somewhat better wages and greater choices.

The seventh conclusion is that the last two decades' expansion of services in public-sector production, utilities and departmental agencies have generated more opportunities of employment for women than the private sectors. While the development drive has in many cases aggravated the work and life conditions of the poor, especially the female poor, state involvement in developing productive infrastructure and disseminating inputs, finance and essential services has not only generated jobs for qualified women and thus an incentive to educate girls, but also opened an arena in which the grass-roots organisations of the female working poor are emerging in parts of South Asia to lobby for and secure access to basic services, inputs and protection from exploitation.

NOTES

1 This chapter was submitted in January 1990. Except for the author's later editing, the chapter is unaltered.
2 Between 1960–1 and 1968–9, the percentage of rural people below the minimum (or absolute) poverty line increased in India as a whole from 42 to 53.5 (Ahluwalia 1978, on the basis of NSS data). Across the states, the poverty ratio increased the most in West Bengal, Assam, Karnataka, Bihar and Madhya Pradesh. In almost all of the states, except Karnataka in which the poverty ratio increased the most, the index of consumer price for the rural poor increased by more than 100 per cent (P. Bardhan: 1974: 277–8).
3 The poverty ratio decreased from 41.3 per cent in 1973–4 to 30.3 per cent in 1983 for India as a whole. The poverty line was based on the officially accepted minimum calories requirement.
4 Kakwani and Subbarao (1989), on the basis of NSS data and a poverty line that is higher than the one used by the Indian Planning Commission.
5 In 1960–1, the states with the worst poverty were Orissa, Tamil Nadu, Kerala and Andhra Pradesh. In 1968–9, the states with the worst poverty were West Bengal, Kerala, Tamil Nadu and Orissa. (Ahluwalia 1978). In 1973–4, the highest incidence of poverty was in West Bengal, Orissa, Bihar and Madhya Pradesh. In 1983, the highest incidence of poverty was in Bihar, Madhya Pradesh, Uttar Pradesh and Orissa. (Ahluwalia 1978; Kakwani and Subbarao 1989; Centre for Monitoring Indian Economy (CMIE) 1987, 1988).
6 Those employed for at least 183 person days in the previous year.
7 The female worker-population percentage in rural India was 31.4 in 1961, 15.5 in 1971 and 23.2 in 1981, with main and marginal workers combined.
8 The average annual rate of growth of per capita state income between 1970–1 and

1984–5 at constant (1970–1) prices was below the all-India average (1.5 per cent) in Kerala, Bihar and West Bengal as well as in the other set of Gujarat, Madhya Pradesh and especially Rajasthan where it was negative (CMIE, state volume 1988, Table 14.1).

9 Many illiterate women working as agricultural or construction labourers or domestic servants want their daughters to be spared this kind of worklife and become either housewives staying at home or have some education for better jobs. To that end, they often want them to have some education, to the extent feasible at their low income level, in order for them to have at least a suitable marriage if not also a better kind of work.

10 National Sample Survey, *Sarvekshana*, April 1988, Tables 13 and 13.1.

11 In the female population 15 years and older, the literacy rate in 1981 was 71 per cent in Kerala and 26 per cent in India as a whole. Lagging well behind Kerala, the literacy rate was roughly a third on the high side in Tamil Nadu, Gujarat, Maharashtra, Punjab and West Bengal. It was less than a fifth on the low side in Bihar, Uttar Pradesh, Rajasthan and Madhya Pradesh.

India
J 16
J 21
R 23

2

GENDER AND CASTE INEQUITIES IN WORKFORCE PARTICIPATION IN URBAN INDIA

A sociospatial interpretation

Saraswati Raju[1]

The generally low level of participation in the labour force in India by urban women and the highly uneven geographical pattern of that participation have been the objects of considerable discussion lately. The tendency in the existing literature on the theme has been either to base the analyses on grossly aggregated data at state levels (D'Souza 1959, 1969; Nath 1970; Mukerjee 1971; Gulati 1975a and 1975b) which tend to obscure sub-regional variations resulting from highly localised historical and cultural conditions, or to study special cases, with little attempt to coordinate these two approaches.[2] In addition, one of the unfortunate practices is to treat the pattern of female employment as existing independently of the male conditions, so that women are portrayed as a separate homogeneous body rather than as an integral part of a larger societal structure. This approach has marred the usefulness of the report on the status of women (Government of India 1974), an otherwise pioneering treatment of the subject which remains to date a baseline reference point for all future discussion on women in India.

THE PRESENT STUDY

Based on 1981 and 1991 census data,[3] this chapter seeks to address the question of variation in the levels of female labour participation in cities with more than 100,000 population.[4] The purpose is to see if the pattern of female labour force participation (hereinafter referred to as FLFP) in cities conforms to the classical contrast in India between the north and the south, or they have distinct urban ambience largely independent of region-specific sociocultural constraints. The detailed analysis follows, but it is interesting to note in passing what my earlier research shows: in India, when the rates of urban FLFP are correlated with the rates of labour participation of rural females, the correlation between the two is positive and highly significant as

74

against the corresponding correlation for male workers (Raju 1984, 1987). Though no causal relationship is envisaged, the existence of a common geographic pattern of variance is implied despite a large difference in rural and urban conditions, and actual labour participation rates. I argue, therefore, that both rural and urban women respond to the labour market more favourably when the atmosphere is conducive. Now this 'conducive' atmosphere has to be much more than just the agricultural context which has often been offered as an explanatory variable to account for variation in the levels of female participation in the labour force (Boserup 1970)[5] because agricultural activity *per se* is not an important component of the urban sector of employment. As such, in India, variation in FLFP is to be seen as one form of expression of the presence or lack of reticence to female involvement in the public sphere outside the familial domain articulated through a complex interplay of region-specific sociocultural constraints posed not only on employment generation, but also the types of jobs which are considered to be 'suitable' for women (Agarwal 1989: WS 53). These constraints transcend rural–urban dichotomy and pervades through the region.

In this study, an attempt is made to analyse the FLFP relative to the corresponding level of male participation with a view to placing the FLFP in the wider perspective and also to expose latent relationships between the genders that often remain concealed when females are studied in isolation from rest of the society.

A number of socioeconomic correlates which contribute to the variation in FLFP such as the presence of those communities which may encourage female work in the public sphere or vice versa and those occupational avenues which may be absorbing female labour in cities are identified.

Although all urban settlements with more than 100,000 population are designated as class I cities in the Indian census, within themselves the range is quite wide, some cities approaching a million plus population status. To capture the size-class variation in the labour force participation and its classification in various industrial categories, the cities are further divided into:

a) million plus (metropolitan cities)[6]
b) 999,999 to 500,000 (large cities)
c) 499,999 to 200,000 (intermediate cities), and
d) 199,999 to 100,000 (small cities).

The basic contention that runs throughout the chapter is that in the Indian context, active participation of females in the labour force is not always related to economic motivations alone. Socially enforced deprivation due to a given social context remains vitally important.

The entire question is found to be extremely complex, and some of the tentative conclusions based on census data need to be substantiated and verified through field observations.

BACKGROUND INFORMATION

Even a cursory examination of the workers' data in the Indian census brings out quite distinctly the abysmally low level of FLFP.[7] Both male and female participation rates are lower for urban populations than their rural counterparts (49.06 and 8.31 per cent respectively for urban India as against 53.77 and 23.06 per cent respectively for rural India in 1981; the respective figures for 1991 are 48.92 and 9.19 per cent, and 52.58 and 26.79 per cent).[8] In the cities, the percentage of workers is still lower, i.e., 47.23 per cent for male and 7.07 per cent for female workers in 1981, and 47.58 per cent and 7.62 per cent in 1991.[9]

These averages, however, obliterate the vast differences that exist between cities, e.g., Valparai in Tamilnadu with 45.00 and 48.39 per cent in 1981 and 1991 in contrast with about 2 per cent in Arrah (Bihar) in 1981 and 1991 respectively.[10] It may, however, be noted that Valparai and Mangalore (in Karnataka with second highest rates of FLFP with 27.57 and 28.55 per cent in 1981 and 1991 respectively) are exceptions.[11] Most of the cities fall in the intermediate category (Figures 2.1 and 2.2).

Even as the overall rates of FLFP are very low in India, they are still lower in the urban economy (Standing 1978: 15). Male participation, by contrast, varies little (Table 2.1).

VARIABLES AFFECTING URBAN FEMALE LABOUR PARTICIPATION

FLFP in urban areas is low because instead of the traditional division of labour in rural areas which is usually complementary among family members, labour relations here become competitive between individual units of labour. Moreover, the changed technologically advanced urban settings require the acquisition of new skills which women usually do not possess (World Bank 1990a). It is, therefore, not surprising to observe that in cities, literacy among women (implying some skill) has a positive bearing upon their participation in the labour force (Table 2.2).

First of all, it is the lack of appropriate skills and education that limits women's employment opportunities in cities; second, social stigma is often attached to women working outside their homes and familial context. On the basis of her observations in Delhi slums, Singh (1978: 79) concludes, 'it can be seen that while the urban occupational structure offers highly limited employment opportunities to women, caste and regional values function to limit their options still further'. This is reflected in the coefficients of variation which are everywhere higher for females as compared to males (Table 2.1).

As discussed earlier in the 'Introduction', females work primarily in those occupations in which household responsibilities can be combined with productive work. This permeability between work and household

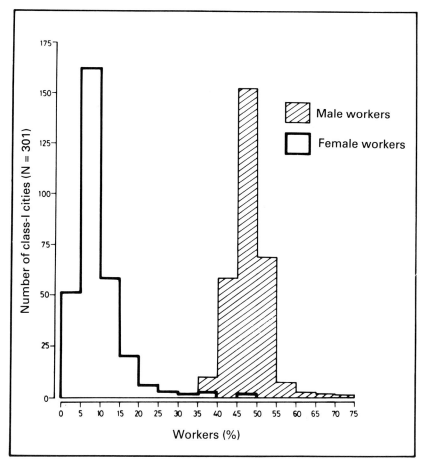

Figure 2.1 Male and female workers in class I cities, 1991
(frequency distribution)

'non-work' – between the inside and the outside – that exists in the rural environment is less evident and boundaries are more clearly defined in the market-oriented urban economy. Moreover, the part-time and seasonal character of agricultural work suitable for women which is available in villages is nowhere matched in urban employment, except perhaps in some cottage industries. Consequent upon such observations, it may be argued notes that the wide gap between the employment rates of rural and urban women is because of unavailability of employment opportunities for urban women.

Part of the variation in the levels of FLFP can thus be explained by the occupational structure of a city and the extent to which it provides activities that are 'appropriate' for women. Though some tendency to seclude women prevails, 'suitability' of occupations may be defined differently in different

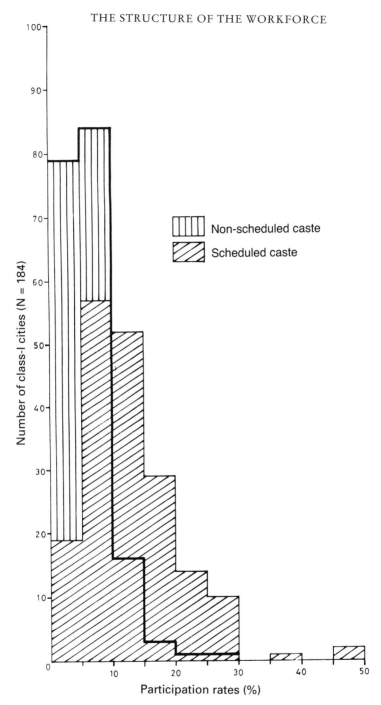

Figure 2.2 Non-scheduled and scheduled caste female workers in class I cities, 1981 (frequency distribution)

Table 2.1 Participation rates in class I cities of India in selected states

States	Male workers		Coefficent of variation, MW		Female workers		Coefficient of variation, FW	
	1981	1991	1981	1991	1981	1991	1981	1991
Uttar Pradesh	46.88	46.02	33.05	5.25	3.12	3.69	48.13	40.37
Bihar	43.01	41.26	7.31	5.26	3.14	3.97	31.48	26.38
Maharashtra	47.54	50.93	11.46	10.14	7.53	9.87	34.89	31.59
Andhra Pradesh	49.28	48.37	6.09	6.53	9.56	9.24	33.95	33.60
Tamil Nadu	52.25	52.66	9.66	9.14	12.35	10.38	68.11	64.13
All India	47.23	47.58	18.50	9.24	7.07	7.62	67.46	63.45

Source: Computed from data, Census of India 1981, Series I, India, Part III-A (i) *General Economic Tables*, Delhi: Controller of Publications, 1987 and Census of India, 1991 Paper-3.

Table 2.2 Correlations of female work participation with selected variables, 1991

Variables	Metro-politan cities (N = 23)	Large cities (N = 32)	Inter-mediate cities (N = 79)	Small cities (N = 67)	All cities (N = 301)
Male workers	–	–	0.27	0.35	0.30
Female marginal workers	0.63	–	0.40	0.40	0.39
Female literacy	0.59	–	0.44	0.29	0.33
Female household industry	–	0.63	0.48	0.37	0.40
Female non-household industry	0.83	0.67	–	0.60	0.64
Female construction workers	0.69	0.56	–	0.26	0.35
Female trade and commerce	0.91	0.83	–	0.50	0.57
Scheduled caste sex ratio	0.74	0.68	0.41	0.19	0.28
Female Christian population	0.72	0.38	0.48	0.39	0.39

Source: Unpublished data on ninefold industrial categories for main workers for the census year 1991 made available in 1993 by the Office of the Registrar General and Census Commissioner, New Delhi, India. Also, Census of India 1981, *Household Population by Religion of Head of Household*, Series I India, Paper 4 of 1984, New Delhi: Controller of Publications, 1987.

Note: Only those results are presented which are statistically significant.

regions (Sopher 1980: 188, footnote 108; Banerjee 1985: 152; Papola 1986: 192–3).

The other possible argument often given as an explanation of regional variation in the female labour force is based on certain sociological premises.

D'Souza (1959: 322–47, 1969: 443–57) has argued that with improvement in the socioeconomic status of a community, participation of women in the

79

labour force tends to decline. His thesis appears to be basically sound, but this particular phenomenon of decreasing female participation in the labour force with increase in status is not uniformly observed.[12] For example, in the present study, the correlation between the percentage of male workers in non-primary occupations (taken, in the absence of a better indicator, as a somewhat arbitrary surrogate variable for socioeconomic development), and the participation rates of women has a negative association when all cities are considered together ($r = -0.26$, significant at 0.01 level). In smaller cities, i.e., with a population between 100,000 to 199,999, this relationship gets further strengthened ($r = -0.37$, significant at 0.001 level). However, this relationship does not hold for metropolitan and other big cities. One may speculate that in such centres, constraints posed by skill acquisition and educational attainments outweigh concerns for status inconsistencies.

If FLFP is seen as an expression of societal approval or disapproval of women in the public domain, one which does not change drastically from rural to urban locale in a given region (as argued in this chapter), then it is imperative that the observed pattern results from long historical processes which involve sociocultural factors indirectly impinging on the prevailing (economic) ideology. Caste is one such factor.

THE CASTE FACTOR[13]

It is well known that in India the employment problems of women of high caste (and class), especially in urban areas, are different from those of the lower castes (and class) (Government of India 1974; Desai 1975; Das 1976). The latter are not as inhibited about engaging in public activity as are women of higher caste (and social class) partly because of their active participation in gainful activities and partly because of the relaxation of the social taboos often still experienced by women of high caste and social status. It has been pointed out that this relationship is somewhat circular, that is to say, the active contribution of females to the household economy results in a more liberal attitude towards female entry into the public domain which may in turn gradually create an environment conducive to female employment.

Following this social reality, the varying presence of scheduled castes has been suggested as an explanation of variation in female employment by some writers (Rao 1978; Sahoo and Mahanty 1978: 331–2; Tripathi 1978; Grover and Krishnappa 1985: 431). However, when Leela Gulati, in her much referred to study on regional variation in female participation rates, tried to establish such a relationship, she found no correlation at the state level, and concluded that 'inter-state difference in the proportion of population accounted for by scheduled castes . . . does not seem enough to explain the inter-state disparities in female work participation' (1975a: 41). I am dubious about the very idea of analysing participation rates at such an aggregate level as the state in India. Moreover, the relationship is not as simple as it was originally thought to be by Gulati. As becomes evident in the analysis that follows, the scheduled castes do not operate in a fixed fashion in isolation from the rest of society as they conform broadly to the behavioural patterns of FLFP in a given region (Das 1976: 130). The differences in factors that influence these patterns tend to cut across caste differences (Miller 1981; Sopher 1980 *passim*). In this context Trivedi's observation is apt:

> The conditions of existence and overall situation of scheduled castes also vary according to the general characteristics of the dominant castes in each region. And the treatment that the scheduled caste women receive from the Muslims, Sikhs, Hindus, Parsees or other minority groups is not likely to be the same in different zones and ways of life. Moreover, the status of scheduled caste women in rural or urban areas is likely to be, directly or indirectly, related to the status of women in other castes and communities of the society.
>
> (1976: 22).

It is true that within the broader framework of regional constraints the scheduled caste women enjoy greater autonomy than women of high castes in terms of their access to work outside the familial domain. The relationship is interesting. Whereas the occupational structure of scheduled castes is governed more by job availability/non-availability than sex discrimination, the non-scheduled caste women with their lack of appropriate skills and formal education find themselves unfit for the employment structure available at large urban centres. Their options are further reduced by their non-scheduled caste status whereby they can join only certain 'respectable' occupations such as teaching, medicine and other white-collar jobs. As such the absolute level of labour participation by scheduled caste women is definitely higher than that for non-scheduled caste women. However, both the levels move in the same direction and the correlation between the two is positive and significant, an association further probed as the discussion develops.

MUSLIMS IN THE CITY

It has been repeatedly argued, and convincingly so, that social and economic structures affect Muslim women much the same way they do other women in a particular region, community and professional groups, especially with regard to attitudes towards women's roles and employment (Mujeeb 1972; Ahmad 1976; Lateef 1990: 218). Aizaz, in her study of Meo Muslims observes that in terms of child marriage, they very closely follow the cultural traditions of Rajasthan where they are located. An interesting aspect is that the Muslims may in fact be converts from caste Hindus (Aizaz 1989).

However, as Lateef (1990) contends, a shift from rural to urban areas

Table 2.3 Temporal trend in sex ratio in class I cities of India

City	1961	1971	1981	1991
Metro	737	779	830	869
Large	813	860	881	876
Intermediate	852	877	878	905
Small	870	888	900	913
All	803	825	846	884

Source: General Population Tables Part II A, 1961 and 1971. See also source: Table 2.1

induces constant changes in group behaviour which tries to approximate the behaviour of the dominant group of a given region by way of absorption of local customs, norms and values. At the same time, group identities and differentiation are constantly maintained (Dubey 1969; Mines 1975; Rothermund 1975). Muslim women (as is the case with scheduled caste and non-scheduled caste women also) have dual membership: one as Muslims and the other as a woman. The already existing tendency of secluding women from the public domain in a given situation may thus be reinforced following the Islamic traditions and the two identities intertwine to depress further their participation in the labour force outside the familial domain. In an earlier study, an intriguing suggestion of relationship seemed to emerge whereby districts in northern India with large Muslim minorities and lingering traces of *ashraf* or 'noble' Muslim élite culture exhibited unusually high disparity in female employment between scheduled castes and the non-scheduled population lending credence to the proposition (Raju 1981: 18).

SEX RATIO

India is one of the few countries in the world where the sex ratio (defined here as number of females per 1,000 males) is declining. In 1981 it was 934 females per 1,000 males reduced to 927 per 1,000 males in 1991. However, as compared to rural areas which follow the national pattern, i.e., 952 females per 1,000 males in 1981 to 939 females per 1,000 males in 1991, the urban sex ratio shows an increase over the decade as 880 females per 1,000 males in 1981 are now 894 per 1,000 males. It may be pointed out that this pattern is a continuation of the earlier decade of 1961–71 which had also registered a decline in the sex ratio in rural areas accompanied by a substantial increase in the urban areas. The class I cities replicate this trend. Table 2.3 shows that although the urban-oriented migration continues to be male selective in nature, increasingly so from small urban centres to metropolitan cities, over the decades, the number of females who join their male counterparts has been increasing. In this context the recent tendency towards family migration, particularly in case of moves involving longer distances, assumes importance (Premi 1990: 193).

Admittedly, this by itself does not necessarily imply a more active work-force participation by the migrant females to the cities. However, it can possibly be argued that increasingly higher sex ratios in the urban population are indicative of the gradual opening up of urban horizons for females and that some indirect inferences linking sex ratios and work status of females can be drawn.

The Indian census collected data on 'reasons to migrate' for the first time in 1981. Although 'marriage' remains the overwhelming reason for females to move, it is important to note that with increase in distance, employment and education become increasingly important reasons for females to move in all types of migration streams (Premi 1990: 200–1). In addition, those females who migrate longer distances for employment usually attain literacy and a certain level of education before moving. It is interesting to note that in proposing a positive link between increasing distance with better education, Premi found no such relationship in the case of male migrants whereas among female migrants, the proportion of those with high school education, diploma holders and college graduates increased significantly with increasing distance (Premi and Tom 1985; also Singh 1984). However limited, these associations imply a certain degree of empowerment for literate and educated females so as to defy and question some of the role models and gender ideology regarding their spatial mobility.

It has been suggested that a sex ratio favourable to females indicates their social status which is further believed to result in their high contribution to the labour market (Grover and Krishnappa 1985: 431). In the present study, also, at an aggregate level this association emerges whereby sex ratio exhibits a significantly positive high correlation with female work participation which is retained even as the cities are grouped in different sizes. However, the proposed association between sex ratios and work participation is not as straightforward as has been envisaged by Grover and Krishnappa and has an element of circularity. That is, high sex ratio may in fact be seen as an input in the labour market or as its consequence. Active contribution of females to the labour market may depend upon an environment conducive to their enhanced social status. In such cases active contribution to work and conducive environment would be mutually reinforcing. Notwithstanding this, the positive association between sex ratio and work participation may be viewed as an encouraging trend. It may imply that more and more women are moving to cities for jobs. A related and crucial point is to identify those occupational avenues which absorb women more readily.

THE GEOGRAPHIC PATTERN OF FEMALE WORK PARTICIPATION

In India, there exists a general gradient of decreasing female employment from north to south (Table 2.1 and Figure 2.3). Broadly, the north Indian

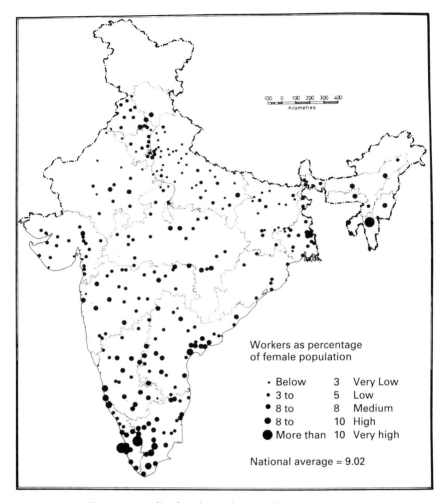

Figure 2.3 India: female workers in class I cities, 1991

pattern of female work participation resembles that of West Asian and North African Arab countries, whereas central and southern India have a pattern more like that of Southeast Asian countries. The urban situation is no exception (Raju 1980; 1982). This has remained remarkably consistent despite changes in the actual rates of female workforce participation.[14]

As just mentioned, the scheduled caste females are not as inhibited about engaging in public activity as are women of higher castes. In the present study, at the aggregate level there exists a positive correlation between female workers and scheduled caste population ($r = 0.20$). This relationship gets strengthened when the female workers are correlated with the scheduled caste female population ($r = 0.21$). Both the values are significant at 0.05

level. However, this relationship does not hold when cities are disaggregated in various size classes. In fact, except in cities with a population of 100,000 to 199,999, in all other categories the scheduled caste population in general and scheduled caste female population in particular have negative bearing upon female work participation, though the correlations are not statistically significant. The male workers do not have any relationship with the scheduled caste presence. It may be noted in passing that whatever weak association existed in 1981 disappears in 1991. On the other hand, as discussed later, the scheduled caste female workers are highly correlated with the non-scheduled caste female workers. The fact that the scheduled caste females do not function in contextual isolation and conform broadly to the behavioural patterns of the region where they are found gets amply demonstrated. As such, instead of expecting a linear relationship between the proportion of the scheduled population and the proportion of female workers, we need to ask where these low-status workers are located, what are the occupations in which they are engaged, what is the working pattern of other women and why females of low status are working in one area and not in another. For example, the smaller cities do have a positive correlation between scheduled caste population and female workers ($r = 0.39$). Besides, it is true that the participation rate among scheduled caste females is, with a few exceptions, higher than the participation rate of non-scheduled caste females. However, the rates for both the groups are themselves highly correlated ($r = 0.73$), indicating a close spatial covariation of the two rates, although the actual levels involved vary from place to place. As compared to this, the correlation that exists between non-scheduled and scheduled caste male workers is much less significant ($r = 0.27$). On the other hand, the scheduled caste male and female workers show a more significant positive covariation in their participation levels ($r = 0.35$) than the non-scheduled caste male and female workers ($r = 0.26$). A more significant covariance among scheduled caste workers is not difficult to understand in view of their place in the hierarchical structure of Indian society. Scheduled caste men do not differ very much from their women in education or skill attainment. The societal restrictions are, moreover, less severe for their women so that their occupational structure is restricted more by job availability than by sex discrimination. It may be of interest to note that in cities, the disparity in literacy between male and female components of population is low for both non-scheduled and scheduled castes. However, a corresponding lowering of disparity in literacy between non-scheduled and scheduled castes does not occur (Kundu and Rao 1986: 446).

The high spatial covariation between non-scheduled and scheduled caste female work participation rates is accompanied by an almost identical rate of increase in the levels of participation for both the groups as is clear from the scatter diagram where the distribution follows the line of parity in a parallel manner (Figure 2.4).[15] It may be recalled that in some circumstances, the caste disparity increases as the participation rates rise. That is, the gap between the non-scheduled and scheduled caste women workers widens in favour of scheduled caste women workers (Raju 1984: 4, 1987: 257). This relationship is also reflected through a negative correlation between caste disparity in female employment and the rates of non-scheduled and scheduled caste participation in the labour force ($r = -0.35$ and -0.39 respectively, both statistically significant). That is to say, with the increase in level of employment for both the groups, the disparity becomes increasingly negative. The way disparity is computed, negative values denote low relative participation by non-scheduled females compared to scheduled caste females. Since the negative correlation mentioned above is slightly more in the case of scheduled caste women, it may be inferred that with the increase in the absolute rates of female participation it is the scheduled caste women whose participation in the workforce increases albeit only marginally. This internal structure is related to economic reality. Within the limitations imposed by society, scheduled caste women are definitely more in need of work.

SEX AND CASTE DISPARITY IN LABOUR PARTICIPATION[16]

So far the discussion has essentially been confined to female workers independent of their male counterparts. Inclusion of the male component of workers, however, provides a perspective whereby the observed patterns of FLFP can be viewed as a part of a larger societal realm. For example, it is often argued that the overall decline of female labour, as well as in certain occupational categories, is not an isolated event, but male workers have also been affected in a similar manner (Swaminathan 1975; Banerjee 1985: 14). The present study also brings out some subtle and intriguing associations when the FLFP is considered in relation to male participation (Figure 2.5).

The proportions of non-scheduled and scheduled caste workers in the respective female population are significantly correlated with the proportions of non-scheduled and scheduled caste workers in respective male populations ($r = 0.26$ and 0.35 respectively). However, the male–female disparity, i.e., sex disparity, within these two groups, covary very strongly ($r = 0.77$). This is not surprising, because what is being demonstrated is the same strong class relations among female workers. As the level of male participation in the labour force does not vary much (CV $= 18.50$ and 9.24 as compared to female workers where the value is 67.46 and 63.45 in 1981 and 1991 respectively), the main contributing factor in the sex disparity among non-scheduled and scheduled castes is the female components of the matrix, i.e., non-scheduled and scheduled caste female workers. These work rates have to be fairly closely related in order to result in a highly associated

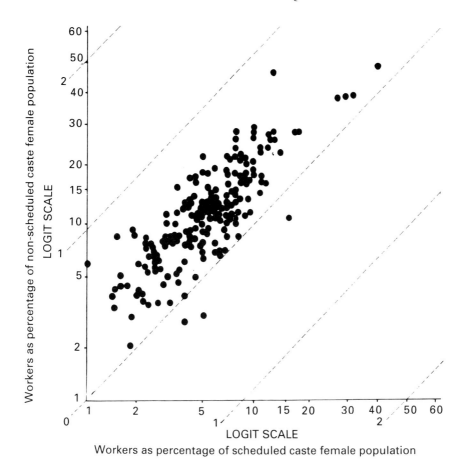

Figure 2.4 Non-scheduled and scheduled caste female work participation rates, 1981 (scatter diagram)

pattern of variation as suggested by the high positive correlation between female workers which, as indicated earlier, is the case. Sex disparity, then, actually involves the caste relationship of female workers. As indicated earlier, this does not mean identical FLFP rates for the scheduled caste and non-scheduled caste women (Figure 2.6).

In sum, the situation is such that scheduled caste male and female workers form a more homogeneous group. In the absence of data for 1991, it is not possible to comment directly on later happenings, but this association is articulated through another very interesting observation. It may be recalled that in 1991, scheduled castes in general and scheduled caste females in particular do not have a bearing upon the FLFP rates. However, they have statistically significant positive correlations with the scheduled caste sex

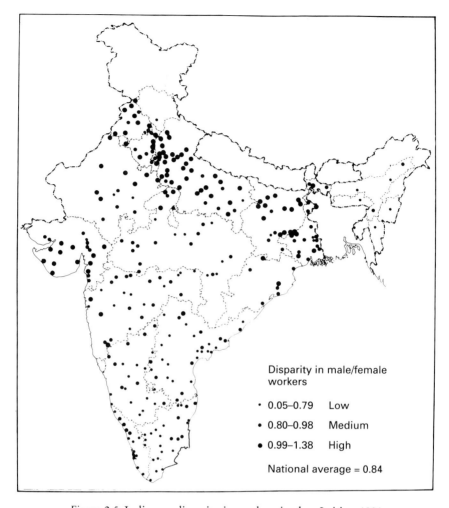

Figure 2.5 India: sex disparity in workers in class I cities, 1991

ratios at every level (Table 2.2). That is to say, scheduled caste females by
themselves do not have any bearing upon FLFP, but in the company of their
male counterparts (conveyed through sex ratio which, it may be reminded,
is defined in this study as the number of females per 1,000 males), their
participation in the labour force increases. This is because their work spheres
are usually not as mutually exclusive as is the case with non-scheduled caste
workers who seem to coalesce more with the scheduled caste female workers
than their male counterparts. This appears to be a far-fetched argument, but
once we consider the attainments in education and skill of women in general
and that of scheduled caste male workers in particular, the levels do not vary
much. The social restrictions are, moreover, less severe for scheduled caste

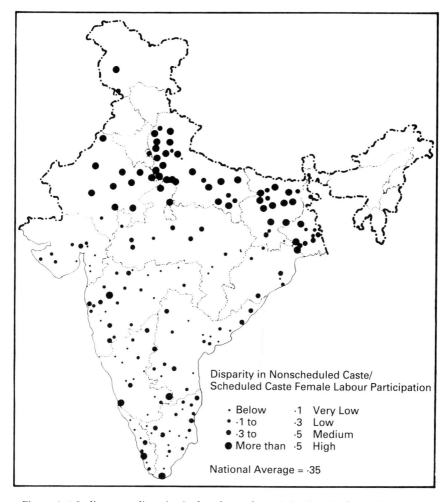

Figure 2.6 India: caste disparity in female work participation in class I cities, 1981

women so that their occupational structure is governed more by job availability than by sex discrimination. On the other hand, the non-scheduled
caste women work in those occupations which are socially 'approved'.
Within the social constraints only those women who do not have options
work. The fact that more than 50 per cent of female workers in urban India
are reported as illiterates in the 1981 census helps us to envisage the types of
work in which women are primarily engaged.

When the non-scheduled caste and scheduled caste female workers are
said to be coalescing, they do so in terms of levels only rather than the actual
nature of the work. Also, the gender relations among the workers vary
across the different size-class cities.

Table 2.4 Workforce and its variation according to size of the city

City	MW^a 81	MW^a 91	FW^b 81	FW^b 91	MW^c 81	MW^c 91	FW^d 81	FW^d 91
Metro	48.60	49.69	6.07	7.30	5.74	8.17	36.00	36.00
Large	46.02	47.67	8.51	6.64	9.82	8.79	48.82	49.82
Intermediate	47.38	47.45	7.15	8.00	9.41	10.49	65.00	57.86
Small	47.32	47.35	7.23	7.93	9.51	8.67	75.79	69.28
All	47.18	47.58	6.83	7.62	9.35	9.24	67.69	63.45

Source: See Table 2.1.

Notes: a Male workers as percentage of male population.
 b Female workers as percentage of female population.
 c Coefficient of variation of male workers.
 d Coefficient of variation of female workers.

CITY SIZE AND LABOUR PARTICIPATION

The male participation rate in the workforce in the cities tends to go down with decrease in city size. This is because of higher employment opportunities in the large cities. This is also because of the shift from part-time underemployed work in agriculture and allied activities in smaller cities (which is not captured in full due to the manner in which workers are counted in the Indian census, i.e. paid work only) to nearly full-time non-agricultural jobs in large urban settlements. For precisely the same reasons, albeit in the reverse direction (with the exception of large cities in 1991) the female labour force increases with the decrease in size of the city (Table 2.4).

It is interesting to note that there exists a positive correlation between female and male workers in cities with the exception of metropolitan and large urban centres (Table 2.2). The absence of correlation between these two variables in the latter urban settlements may be due to more clearly defined and segmented work spheres for male and female workers there as compared to smaller cities. This may also be taken to imply a relatively more competitive environment in larger urban centres as compared to smaller ones where the labour-market relations would usually be complementary among male and female workers, a proposition which requires further probing before any definite statement can be made.

It has been observed that with increasing size of a city, the percentage of workers in non-household manufacturing increases (Table 2.5). Such a predominantly manufacturing nature of large-size urban settlements has been a consistent pattern during the 1960s as observed by Bhalla and Kundu (1982: 62–3). Their remark is directed at class I cities in general and may be modified in the context of the present study. That is, within class I cities, it is the metropolitan cities which exhibit this characteristic the most.

The distribution of household manufacturing in different size-class of

Table 2.5 Workers in industrial categories according to the size of the city, 1991

City	Primary		Secondary						Tertiary[b]	
			Va		*Vb* Non-		*VI*			
			Household industry		household industry		Construction			
	M	F	M	F	M	F	M	F	M	F
Metro	2.22	4.42	1.83	3.76	35.51	19.98	6.46	5.60	53.98	60.24
Large	11.29	15.18	2.15	7.65	27.95	19.77	5.61	4.82	53.30	52.50
Intermediate	10.94	15.89	2.11	12.19	26.81	24.24	5.75	5.61	54.39	43.99
Small	16.35	27.64	4.17	11.80	20.37	17.08	5.31	2.98	54.80	40.51
All	7.65	12.87	2.35	7.80	29.89	20.35	5.98	4.54	54.12	54.24

Source: Unpublished data on ninefold industrial categories for main workers for the census year 1991 made available in 1993 by the Office of the Registrar General and Census Commissioner, New Delhi, India.

Note: Primary consists of industrial categories: I cultivators, II agricultural labourers, III livestock, forestry, fishing, etc. and IV mining and quarrying. Tertiary pertains of VII trade and commerce, VIII transport, storage, communication and IX other workers.

cities shows a very small component of the household sector and in the case of metropolitan cities the share of male/female workforce in this sector is about 3 per cent only. Moreover, a more developed manufacturing sector, in general, is associated with a declining household sector. This relationship was again strengthened at the metropolitan level where workers in the household sector were negatively correlated with those in the manufacturing sector. This was true for both male and female workers in 1981 ($r = -0.55$ and -0.41). However, in 1991, this negative relationship has weakened in the case of male workers in cities of all sizes although the correlation values are statistically insignificant. For female workers, the negative relationship observed between the household and non-household workers in 1981 actually disappears to be replaced by a positive association which is statistically significant when all cities are taken together. The metropolitan cities are exceptions. However, for large cities the association is statistically significant ($r = 0.72$). What these associations suggest is what has already been discussed in greater detail in the introduction: the gradual obliteration of boundaries between household and non-household sectors coupled with an increasing mutual inter-dependence between the two whereby much of the work (which ultimately becomes an input in non-household factory industry) is being done by female workers at home on a contractual basis, particularly so in large urban centres. At another level, the importance of the manufacturing sector in the employment of women becomes evident. As the proportion of female workers in the population increases, their share in manufacturing increases in a positive and significant way (Table 2.2). For male workers, no such association exists.

91

Table 2.6 Stepwise regression of female work participation on selected variables (all cities)

Independent variable /	Dependent variable / % of total workers/population	Per cent variation explained[a]
1	Sex ratio	13.86 (+)[*,b]
2	Male non-primary workers	22.90 (−)
3	Female workers (other than household)	27.47 (+) [*]
4	Female Christian population	30.25 (+) [*]
5	Female scheduled caste population	35.24 (+) [**]
6	Female workers (household)	36.43 (+) [**]
7	Population size	36.73 (+)

Source: See Tables 2.1 and 2.2.

Notes: a The mathematical sign in the parentheses indicates the slope of the b-coefficient.
   b Level of significance: [*]1 per cent, [**]5 per cent.

Larger urban centres thus do not appear very congenial for the absorption of female labour as a whole. However, once in the cities, the non-household sector opens up avenues for them (Banerjee 1985: 19).[17] Trade and commerce, and construction are other venues which emerge as important components in 1991 for female workers as their share in these activities has a positive bearing upon their proportion as workers in the respective population. It is to be noted that in the earlier decades, these industrial categories did not emerge as very important arenas for female workers.

THE REGRESSION MODEL

Consequent upon the discussion so far, several variables may be identified and their composite effect on FLFP can be evaluated. However, such an evaluation is highly constrained by the availability of published data especially when the analysis is carried out at a pan-Indian level, and the complexity of the questions raised involve a closer scrutiny of agro-ecological, cultural, historical, structural and social factors which are not easily quantifiable and which the official data fail to capture. Notwithstanding these limitations, some results of stepwise regression for different size-class cities for 1981 are presented in Tables 2.6, 2.7, 2.8, 2.9 and 2.10.

At the risk of repetition, a brief explanation is in order. It is important that the hypothesised relationships be viewed strictly in conjunction with earlier observations in this chapter. Thus, the percentage of Muslims in the population is expected to suppress the FLFP. In contrast, the presence of

Table 2.7 Stepwise regression of female work participation on selected variables (metropolitan cities)

Independent variable / Dependent variable / % of total workers/population		Per cent variation explained[a]
1	Female Christian population	36.93 (+)***[b]
2	Female scheduled caste population	40.85 (+)***
3	Female workers (other than household)	43.54 (+)*
4	Female workers (household)	47.86 (+)**
5	Sex ratio	50.86 (+)*

Source: See, Tables 2.1 and 2.2.
Notes: a   The mathematical sign in the parentheses indicates the slope of the b- coefficient.
       b   Level of significance: *1 per cent, **5 per cent, ***10 per cent.

Table 2.8 Stepwise regression of female work participation on selected variables (large cities)

Independent variable / Dependent variable / % of total workers/population		Per cent variation explained[a]
1	Female workers (other than household)	39.18 (+) *[b]
2	Female Christian population	56.93 (+) **
3	Sex ratio	60.25 (−) **
4	Female workers (household)	63.65 (+) ***
5	Female Muslim population	65.26 (−)
6	City size	65.63 (+)

Source: See Tables 2.1 and 2.2.
Notes: a   The mathematical sign in the parentheses indicates the slope of the b-coefficient.
       b   Level of significance: *1 per cent, **5 per cent, ***10 per cent.

Christians and scheduled castes in the population would, in general, enhance FLFP.

The percentage of workers in non-primary occupation is taken as a somewhat arbitrary measure of the overall level of development and improved socioeconomic status of the population. In the occupational categories, only the household and non-household employment is taken because they represent two distinctly different scenarios in terms of absorption of female labour. Household industries offer a context where women can combine housework with work whereas the factory-based industries, in general, discourage female employment. However, metropolitan cities and

Table 2.9 Stepwise regression of female work participation on selected variables (intermediate cities)

	Independent variable / Dependent variable / % of total workers/population	Per cent variation explained[a]
1	Sex ratio	23.13 (−) **[b]
2	Female Christian population	31.12 (+)
3	Female workers (household)	31.50 (+) **
4	Female scheduled caste	32.28 (−) ***
5	Female workers (other than household)	32.49 (+)
6	Female Muslim	34.10 (−)
7	Male non-primary	36.99 (+)

Source: See Tables 2.1 and 2.2.
Notes: a The mathematical sign in the parentheses indicates the slope of the b-coefficient.
     b Level of significance **5 per cent, ***10 per cent.

Table 2.10 Stepwise regression of female work participation on selected variables (small cities)

	Independent variable / Dependent variable / % of total workers/population	Per cent variation explained[a]
1	Female scheduled caste	15.73 (+)**[b]
2	Sex ratio	27.48 (−)*
3	Male non-primary workers	32.21 (+)*
4	Female Christian population	41.56 (+)*

Source: See Tables 2.1 and 2.2.
Notes: a The mathematical sign in the parentheses indicates the slope of the b-coefficient.
     b Level of significance *1 per cent, **5 per cent.

large urban centres do not have much scope for household industries. The interplay between these varying factors is intriguing and worth analysing. Finally, the effect of the size of cities on female labour is evaluated.

While it is not possible to comment on different categories of size-class individually, several trends may be loosely identified.

With limited variables entering into the stepwise regression model, the percentage of variation in FLFP that is explained is minimum when all cities are taken into consideration (Table 2.6). This is because here the range of population is anywhere from 100,000 to a million plus and the permutation and combinations of various factors differ tremendously for different size-classes of the cities. What is important to note, however, is that the sex ratio has a negative sign implying that at aggregate level, female presence is not always associated with increase in labour participation.

This negative relationship with sex ratio is replicated at all city sizes except at the metropolitan level. The earlier discussion on sex ratio brings out clearly that despite increasingly favourable sex ratios over the decades, the urban settlements remain essentially male-selective in terms of migration flow (Table 2.3). This trend is much more pronounced in the case of metropolitan centres (Bhalla and Kundu 1982: 65). Under such circumstances, it may perhaps be reiterated that although larger urban centres are unsuitable for the absorption of female labour as a whole, once in the metropolis, the women find employment relatively easily compared to smaller urban centres.

This point needs further clarification. Although, as observed earlier, sex ratio *per se* exhibits a positive association with FLFP, the relationship changes in conjunction with other variables. Further, the negative association between sex ratios and female work in cities (other than the metropolitan centres) in the stepwise regression model should not be seen as contrasting with my earlier observation of increasing FLFP with decreasing city size. What is being hinted at is that the FLFP rates lag with increasing sex ratios in urban centres with the only exception being metropolitan cities.

As far as scheduled caste females are concerned, at the aggregate level, for the small and metro cities, their presence increases FLFP rates. However, everywhere else they have a negative bearing upon these rates. This may be because of the occupational structure available in small cities as compared to intermediate and large urban centres. In the absence of relevant information, no definite statement can be made. None the less, it strengthens my earlier observation that although the scheduled caste women almost always have a higher participation rate as compared to non-scheduled caste women, their proportion in the population *cannot* be a predictor of the level of FLFP.

When the cities are disaggregated, the percentage of variation explained increases considerably especially in case of metropolitan and large cities. The presence of Muslim and Christian females affect the FLFP in the hypothesised direction. However, the household or the non-household composition of the workforce appears to be more important, especially in case of metropolitan cities.

Even with a less than satisfactory data base, in the metropolitan centres and large cities, the occupational structures available seem to have far more significant bearing upon FLFP as compared to small and intermediate cities where caste composition appears to contribute more importantly to the variation in FLFP. Interestingly enough, male workers in non-primary occupations (taken as surrogate of improvement in economic status) do not enter into the regression model for metropolitan centres and large cities. Perhaps in smaller and intermediate cities, on account of their relatively more traditional sociocultural ethos, women's withdrawal from the public domain with increase in family status is maintained.

In sum, the present analysis identified certain subtle geographic variations in FLFP in urban India. A few points have been clarified, and some remain ambiguous; perhaps these can provide certain pointers for further research in this field.

NOTES

1 My thanks are due to O.P. Bohra, National Institute of Public Finance and Policy, New Delhi and Mary Jose, a doctoral student at the Centre for the Study of Regional Development, Jawaharlal Nehru University for their help in computation of data and cartographic assistance respectively at various stages.

2 Most of these cases deal with very modern, industrialised cities like Delhi, Bombay and Chandigarh. Even then, confusingly enough, authors tend to make questionable generalisations on the basis of a single case. The following statement by P.T. Kuriakase in his preface to *The New Breadwinners* (Wadhera 1976: x) provides an example: 'We felt that a study of the situation in Delhi would be sufficiently indicative of the urban situation in general in the rest of the country.'

3 The 1991 data have been used only for total workers. For disaggregated analysis of the scheduled caste and non-scheduled caste population as well as for Muslims and Christians the data used are for 1981 as they are not yet available from the 1991 census. It may, however, be pointed out that despite differences in absolute terms, there has been no substantial redistribution of these segments of the population in 1991.

4 A city in the Indian census is any settlement which has a) a municipal corporation, municipal board, cantonment board or notified area, etc.; b) a minimum population of 100,000 people; c) 75 per cent male workers in non-primary occupation; and d) a density of population of at least 1,000 per sq. mile. Some of these cities are designated as urban agglomeration (UA) which means a) a city or town with contiguous outgrowth, the outgrowth being outside the statutory limits but falling within the boundaries of the adjoining village or villages; b) two or more adjoining towns with their outgrowth(s); or c) a city with one or more adjoining towns with their outgrowths forming a continuous spread.

5 See, for example, the 'Introduction' of this book.

6 This classification of cities on the basis of their population is taken from Premi (1980: 16).

7 It is also possible that poor urban women *are* in fact economically active, but in an informal economy which is captured even less well in official statistics than the rural subsistence economy. Numerous studies carried out in poor urban areas show actual female participation rates of around 40 per cent. Moreover, the few longitudinal studies available show that the labour force participation of poor urban women is increasing much faster than men's (Bapat and Crook 1988). This suggests that problems of measurement and definition may make changes in urban female participation rates suggested by macro-level data less than reliable (World Bank 1991).

8 These are main workers, i.e., those who were engaged in any economically productive activity for 183 days or six months or more during the year preceding the date of enumeration. Those who worked for less than the stipulated period are identified as 'marginal workers'. However, in the urban situation, especially in class I cities, inclusion of marginal workers does not improve the labour participation rates much.

9 It is to be noted though that these are crude rates. The demographic structure of the Indian population containing a large fraction of non-working children greatly depresses the actual level of participation in the labour force. In this study, however, the evidently less satisfactory crude rates are used because age-specific data for workers and the disaggregated subsets of population that are of interest here are not available for cities.

10 Kannappan's extensive analysis of urban employment based on more than seventeen case studies from developing nations clearly identifies the importance of literacy, education and technical training as general attributes purveyed by formal sectors of urban employment. Although he does not disaggregate these attributes into gender components, there is enough empirical evidence to show that females are almost always universally more deprived than their male counterparts in terms of acquisition of these attributes (Kannappan 1985: 714–15). Also, see Papola (1986).

11 Valparai in the Anamalai Hills of Tamil Nadu is the only city with very high functional specialisation in the tea plantation industry (Mitra *et al.* 1981). It covers an area of approximately 394 sq. kms, by far the largest area to be included in any urban settlement. This implies inclusion of large rural pockets within the confines of the city. This phenomenon may account for the extraordinarily high FLFP.

12 For example, Andrea Singh (1980: 80), in a study of women in Delhi's squatter settlements, notes that a significant percentage of working women did *not* want to discontinue work even if their husband's income increased. Moreover, it can be argued that in most cases, an increase in prestige by virtue of the nature of the occupation does not necessarily mean such an increase in income as to allow the wife to withdraw from work.

13 As mentioned in the 'Introduction', the term 'scheduled' refers to historically 'underprivileged', 'deprived' or 'depressed' castes *vis-à-vis* other castes. In this analysis, the scheduled tribes are excluded from all calculations. Thus, the non-scheduled castes are: total − (scheduled castes + scheduled tribes). However, the non-scheduled or higher castes as defined here consist of varying groups living under a wide range of socioeconomic conditions, but the data available in the census do not allow such distinctions to be made.

14 This kind of relative stability in the spatial variation over time has been observed in a rural context (G. Sen 1983) and also with other variables (Raju 1988).

15 Disparity between two groups, of which different proportions possess a particular property (in this case urban labour force participation) is here measured by Sopher's disparity index as modified by Kundu (Sopher 1974, 1980; Kundu and Rao 1986). With axes representing the two proportions drawn on the logit scale lines of equal disparity appear as straight line. The two formulae are as follows:

$$DIS = \log X_2/X_1 + \log (100 - X_1)/(100 - X_2)$$
$$DIK = \log X_2/X_1 + \log (200 - X_1)/(200 - X_2)$$

For a detailed discussion on how disparity index is a better statistical tool see Sopher (1980, especially footnote 87).

16 Non-scheduled caste/scheduled caste disparity is calculated by converting the participation rates of both the groups and also for male and female within the groups into logits and subtracting the scheduled caste values from the non-scheduled caste values; subtracting the values for female participation from the values for male participation yields the male/female disparity in labour participation. The presumption is that the behaviour of the non-scheduled castes and the males, as the case may be, is more representative of a normal situation than the

97

behaviour of the scheduled castes and the females. Therefore, the first group is taken first in the disparity pair. In Kundu's method, it is usually the lower value in the pair which is taken first.

17 Out of twelve metropolitan cities, i.e., population more than 1,000,000 (in 1981), six have manufacturing as their functional specialisation.

3

WOMEN, AGRICULTURE AND THE SEXUAL DIVISION OF LABOUR

A three-state comparison

Joan P. Mencher

This chapter explores women's involvement in rice cultivation in India, focusing on data from two of the southern states (Kerala and Tamil Nadu) and one state in the east (West Bengal), three of India's main rice regions. It is concerned with variation in the sexual division of labour in agriculture, both at the macro (variation from state to state, and from region to region within each state) and the micro level (variation from village to village, and between different castes and socioeconomic groups within the same village). It is based on data collected from a large-scale study of women and agriculture which the author and Dr K. Saradamoni conducted between 1979 and 1982 in ten villages of Tamil Nadu, ten villages in Kerala and eight villages in West Bengal, as well as intensive field work carried out by the author in these areas between 1958 and the mid-1980s.[1]

The chapter also explores the concept of 'hard work' in the context of South Asian agriculture. The question raised here is: how has it come about that women's work is regarded as easier than men's work, both by the workers themselves and by many social scientists who study them? Who is doing the defining and what are the criteria being used? Taking the lead from Bina Agarwal (1988: 532), the chapter looks at the dialectical relationship between the material context of women's relationship to agriculture and the land on the one hand, and gender ideology (here related to the valuation placed on what work they do in agriculture) on the other.

My research in India since 1958 (see especially Mencher and D'Amico 1987, Mencher 1988) has established four essential facts about the roles of female agricultural labourers in India's main rice regions:

a) women perform a very large part of the heavy manual work in rice cultivation in India. This is important to note because as compared to African and Southeast Asian women in agriculture, the involvement of Indian women in actual field work has often been ignored.

99

b) in the case of landless labourers, women's income is a major contri-
    bution to household income – essential for survival – even where wage
    rates for females are lower than for males. This fact cannot be empha-
    sised enough because the conventional development theory often
    assumes that even when Indian women work in agriculture, their
    income is largely supplementary to that of the males in their households.

c) for both males and females, actual participation in manual field work has
    always been culturally considered to be degrading, fit only for a lower
    caste, more so if the individual has the means to pay others for this
    work. While this may vary between community or by caste (e.g. be-
    tween Christians and Nayars in Kerala, or between Rajputs and Jats in
    the Punjab as reported by Sharma 1980: 118), there is no question that in
    India manual work has always been looked down upon. Withdrawal
    from field work has always been associated with higher socioeconomic
    class. It has not, however, always involved withdrawal from all work
    associated with agriculture, since even those who have acquired enough
    land to withdraw from manual work in agriculture still have to carry out
    numerous managerial tasks. For women, withdrawal from field work
    has often merely meant heavier work in the household compound rather
    than leisure, since the size of the harvest is larger with a bigger holding.
    (Women are often in charge of paying workers, getting implements
    ready for them, preparing food for the labourers at noon time, etc.)
    Even where they hire someone to be in charge of these activities, they
    are often called upon to perform other tasks.

d) tasks in field work have always been segregated by gender in South Asia.
    In addition to the many tasks women do in rice cultivation, women may
    also plant vegetables and work the soil manually, but they have only
    rarely been considered worthy to plough the fields.

HISTORICAL BACKGROUND

For at least the past 1,500 years and possibly longer, wet-rice cultivation in
India has been associated with more complex social structures than those
generally found in the traditional wheat areas or areas of lesser grains. Wet
rice produces much higher yields than other grains. Thus, throughout
history, the rice regions tended to have the densest population in rural India.
In fact, the rice regions of Kerala have the highest person–land ratios in rural
India. Since rice cultivation involves a great deal more of unpleasant and
messy work than wheat cultivation, those who could afford it have traditionally used the labour of others, both male and female, to work the land.
Landowners are usually quite explicit in talking about this. People from
low-ranking groups provided most of the actual field labour in these regions,
which tended to develop a three-tiered socioeconomic system consisting of:
a) landowners, b) tenants, and c) agricultural labourers, who in southern

100

India prior to the nineteenth century lived under conditions of rural servitude similar to slavery. (In Bengal, this type of labour was more typically provided by people from tribal groups who migrated in and out of the villages.)

One cannot underestimate the importance of the role of class and land ownership on the sexual division of labour. Landownership gives power to both men and women as agricultural supervisors, though not all women are in a position to exercise this potential power. While members of middle- and upper-class landowning households have always participated in agricultural management, the actual physical work has been done primarily by much poorer, if not landless labourers, whose reliability, physical strength and skill were valued by the employers. For this class of people it was not a question of male vs female in any given locale, but of a person's ability to do the required tasks.

While some landowning men might plough, usually along with teams of hired males, the only other activities in which they participated involved supervision, which could be carried out by men or women while holding an umbrella (as protection against the sun or rain) and wearing clean white clothes. This was particularly true of upper-caste landowning households. Very recently, this has been changing in parts of Kerala, where some educated upper-middle-class landowners are determined to break the unions or prove their independence from labourers by doing as much of the work as possible themselves.

The value system associated with this social setup is considerably different from that described by Maclachlan (1983) for Yavaahalli, the village area which he studied in Karnataka, or for that matter from that which has often been described for many parts of Europe. Specifically, while the ability to do hard physical work was considered desirable in a servant irrespective of gender, such ability was not traditionally highly valued in members of the landowning classes. And even among tenants, both males and females traditionally avoided as far as possible any work which kept them in the hot sun or standing for hours in mud. If they could manage to have others do it for them and still have enough to eat, they did so.

There is some variation by caste in the extent to which this attitude towards physical work prevails. People belonging to castes that traditionally did not own land are perhaps somewhat more willing to continue doing hard work even after they manage to gain possession of land. None-the-less, in the case of both Tamil Nadu and Kerala this attitude towards manual work is at least partly responsible for the tremendous concern among members of landowning groups to get their children educated. A very similar attitude has been noted in West Bengal and in the coastal parts of Orissa by many observers. Thus class has always been a major determinant of the division of labour, and of what constitutes a 'good worker', a 'good man', a 'good woman', etc. While physical fitness is certainly important, supervisory or managerial skills are commonly regarded as superior to brute strength, and these skills have never been regarded as sex-linked. Class more than gender influences cultural values in this situation.

The implications of the above for agriculture are obvious. Any farming household is concerned to get the best possible yields from its land, and certainly even members of landowning castes know that hard work is important; yet different kinds of households employ different strategies to reach this goal. As noted above, some male landowners do some field work (mainly ploughing and levelling). Some landowning women also work hard in their own fields, especially those who belong to lower castes and untouchable or tribal groups, since these women usually directly supervise activities such as transplanting and weeding by working along with hired labour, and sometimes harvest alongside their hired female labourers as well.

As far back as we can find any documentation, women played a major role in the actual field work in these regions, though most of these women came from untouchable or other low castes, or tribal groups. It is not yet entirely clear whether the important role of women as cultivators of rice in these areas from time immemorial is linked to a) the arduous nature of the work; b) the fact that women were the food gatherers in neolithic times (and may have even been the first to domesticate rice, as some have suggested); c) paddy production being so labour intensive; or d) some combination of these factors (Mencher and Saradamoni 1982).

With the transitions that occurred both during the period of colonial domination and in the years since independence, there have been enormous changes in the patterning of agrarian relations in India, nowhere more evident than in the rice regions under consideration in this chapter. In all three states there has been a massive penetration of capital into agriculture. To a large extent, the former bonded or attached labourers have been replaced by wage labourers, who now work as casual labourers for whoever might give them work. Those who work as permanent labourers for particular landowners are more fortunate, since they manage to get more days of work than casual workers (Mencher 1978).

In India, a relatively large surplus labour population of underemployed males does not reduce female participation in rice cultivation. In my earlier work in one of the Palghat villages (also included in the MS project), I found that on an average landowners used women for 417 hours per acre of wetland per crop season whereas they used males for 106 hours per acre (Mencher 1980). These figures vary from region to region, with higher male–female ratios in Tamil Nadu and West Bengal than in Kerala (for reasons to be discussed below). None the less, women play a major role in rice production in all three states. In Chinglepet district of Tamil Nadu, where I worked in 1963 and 1970-2, I found that while men may be involved in a greater number of operations connected with paddy cultivation, there is a greater time-wise demand for women, since they work in some of the most labour-intensive aspects of paddy cultivation. Thus, while it might take two men five days to plough a given field, it might take ten women four days to transplant the same amount of land (Mencher 1985a: 354).

DESCRIPTION OF THE AREAS STUDIED

As noted above, the MS study was carried out in 28 villages, in nine of which the author had lived and worked previously. (She spent considerable time in two of the others during the study period.) Two of the remaining villages had been studied by other western anthropologists, who have published materials on them, and a few others had been studied by Indian social scientists.[2]

An attempt was made to cover each of the main ecological zones in which rice was traditionally a dominant crop. Thus, in Tamil Nadu all the coastal districts with sufficient water to grow paddy, and two selected pockets where rivers (the Thambrapani and the Vaigai) provide the main source of irrigated paddy are included. Coimbatore is not included because despite its recent importance it was not traditionally a rice-growing district. In Kerala our sample covers the entire length of the state including both irrigated and non-irrigated villages, and villages with a widely differing variety of caste and religious groups.

In West Bengal, for logistical reasons, we were obliged to confine the study to eight villages in four districts with no North Bengal districts. However, our sample covers a fairly wide range of the ecological and social conditions found in the state including both irrigated and non-irrigated villages, in the active and moribund delta areas, as well as the forest region.

The present discussion draws on material from all of the villages for part of its analysis, and a smaller sub-sample of villages for other parts, since some material is only available for villages where the author has actually lived, and other material is only available for villages studied during the second part of the field project.

GENDER AND THE DIVISION OF LABOUR IN AGRICULTURE

Manual Work

Work with animals

Much of the early work on the physical basis for a gendered division of labour rests on the empirical observation that, from the earliest introduction of the plough, it was seen that animal-drawn ploughs as well as other work using animal traction have normally been handled only by males throughout Europe, the Middle East and Asia (and later on in the New World). Yet the implications to be drawn from this are far more complex

103

than has often been assumed. Furthermore, agricultural societies using the plough range from those with a great deal of inequality between the sexes, as in South Asia, to those with much greater equality as in Burma or other parts of Southeast Asia.

An often neglected point is that women were not spared work with animals as such, and even in many of the most sex-segregated societies (including many parts of India) women have always been responsible for cleaning the cowsheds, washing the animals, making cowdung cakes (all the messiest and dirtiest of work), milking the cows and often attending to animal deliveries. In some areas washing and milking is done by young males, but this often varies from household to household. In addition, in poor tenant or landowning households without a bullock and unable to hire one – if ploughing has to be done urgently – a wife might, in rare cases, be hitched up to the plough (Madgulkar 1958: 57).

In southern Asia, women have not had access to any labour-saving devices that make use of animal power. Thus, in parts of Kerala where threshing is done by beating the harvested sheaves against a board or some other surface, it is done primarily by women. In Tamil Nadu, when it is done using bullocks it is done exclusively by males, but when it is done by beating both men and women participate. The men tend to do most of the beating, the women serving as helpers to hand them the bundles of sheaves and pick up the scattered paddy.

While working in one Kerala village, I was told about an unusual woman in the village, who many years previously had begun doing her own plough-ing. A unique case, it illustrates some important aspects of the question. She was a widow living in a village I studied in 1971 and again in 1981. (She herself had died in the late 1960s.) While collecting data on the sexual division of labour, I asked if people had ever heard of a woman ploughing. Most of my informants laughed and said no, it was impossible, but one older man told me about the widow who, though she had enough money to hire someone, took it into her head that she was going to do her own ploughing, and started doing it. At first she was unable to do it, but slowly she improved. Even so, it was the general opinion that only a crazy woman would plough – not that she was crazy in other ways, but everyone knows that ploughing is only done by men.

In one of our Kerala villages where our village assistant asked several of our informants about whether women could ever plough, we were given two very interesting answers, one by a Muslim woman, the other by a Hindu. Both these women claimed that because ploughing has to be done with sanctity or by those who are sacred, only men can plough. It is clear that they are not discussing physical strength, but rather the belief that those who menstruate pollute the earth. Among the Oraon tribals of Bihar it is said that:

if a woman were to plough, there would be no rain, and calamity would follow. . . . When women in desperate circumstances have ploughed family land they have been severely punished: a tribal woman in Bihar was yoked to the plough along with a bullock and forced to plough the village headman's field.'

(Dasgupta and Maithi 1980 quoted in Agarwal 1988: 562)

Kishwar writing about Ho tribal women in Bihar notes that if a woman accidentally touches a plough she can be heavily fined by the *panchayat* because her action is believed to bring ill luck to the entire village such as drought. She notes that 'In rare cases, they could even be stoned to death' (Kishwar 1987; also, Sharma 1980: 114).

When the two stories from Kerala were told to a Philippine woman who had worked in rural areas, she pointed out that in the Philippines women often plough, but never pull out the paddy seedlings for transplanting.

However, there is no need to raise a controversy on the question of whether or not women should plough. In view of the class-based hierarchy in India, it seems to me to be reductionist to try to explain the higher status of males (culturally and in terms of power relations) on the basis of their physical strength. The above incidents are mentioned more in order to indicate the cultural basis for practices rather than to present an argument for men or women doing a particular job.

Transplanting and weeding

In many areas where only women are expected to transplant, most men will not do this work even if there is no other work for them. They will explain carefully that only the women know how to do it or that it is 'women's work' or easy work – even though in another region the task might be classed as 'men's work' or as gender neutral. Here, learning and training also play a part.

One obvious difference between Tamil Nadu and Kerala is that throughout Tamil Nadu, men pull the seedlings and women put them in, whereas in Kerala both of these are considered to be female tasks. Pulling is considered to be 'hard work' in Tamil Nadu but 'easy work' in Kerala. Throughout both Kerala and Tamil Nadu untouchable and low-caste women (Hindu, Muslim or Christian) transplant, whereas few of the higher-caste female agricultural labourers transplant, and for those who do so it is considered shameful. In fact, some of these women refused to admit doing such work when they were first interviewed. Many of these higher-caste women were not as good at transplanting as others because they had not learned to do it when they were young, but only turned to it later under economic duress.

Weeding is both castelinked and gender segregated in the south, though not in Bengal. And again, it is linked to lower-caste women, as opposed to harvesting and postharvest work which is more acceptable to the higher-caste agricultural labourer women (presumably because it does not involve standing in mud). In Bengal we found both males and females doing transplanting and weeding, but these tasks were reserved for tribals, semi-tribals and some untouchable castes with no caste Hindu women doing this kind of work.

Other manual tasks in rice cultivation

Regarding the actual tasks performed by males and females in each of the areas covered by the MS study, ploughing and all activities that use bullocks as draft animals are done by males (these account for a very short portion of the agricultural calendar). In contrast, there are very few jobs in areas of paddy cultivation that are not done by women (at least in some districts). Other activities that tend to be restricted to males alone include:

a) climbing trees for coconuts or to tap them for toddy or any activity that raises the body up more than a step or two, such as collecting mangoes. Even little girls are strongly chastised for attempting to climb trees because it might show their genitals. Thus, *even in the folk explanation* culture and not physique is invoked (because clearly it will make no sense to claim that energetic young girls are incapable of climbing trees).

b) most operations connected with irrigation and dewatering are carried out exclusively by men, though sometimes a woman might use the *ettam* (a water lift) in a household compound. Using a water lift can raise her above the ground and potentially expose her (though only if a male went out of his way to crawl on the ground!). The main reasons given for this restriction on women are cultural: the work must often be done at night, it can sometimes lead to physical fights with neighbours, and in the case of waterlifts in fields, might expose them too much. When it comes to pump-sets, cultural explanations also exist, e.g. the danger of a female being raped if she is alone in the fields. Even male landowners themselves usually hire other males to look after pump-sets, because they do not want to get up at night.

c) applying chemical fertilisers, which are light in weight but very expensive, is generally men's work (though Bengal villages have started using female labourers for this).

d) digging is done by men rather than women, though in a few villages women also dig. Most digging is done for *bund* construction and for garden work. Women dig for gardens but not on the *bunds*. On the other hand, women sometimes work with the men fixing the *bunds*.

e) applying pesticides is men's work in all areas. This is not to protect the women from the bad effects of pesticides because they do stand in pesticide-saturated mud while transplanting and weeding. The only explanation given for this is that the 'pesticide costs a lot'. But I suspect there might be other factors involved. It is interesting that in one of our Bengal villages where only tribal people worked in the field, women had started to apply pesticides in 1981.

f) driving and repairing tractors and other machines, as well as bullock carts, is exclusively men's work. In general traditional attitudes towards females appear to have been welded together with western ideas about women and machines to keep women from using them. This is despite there being more women in engineering colleges in India, especially in southern India, than in countries like the United States.

At least in some parts of each of the three states, or among some caste or tribal groups in each state, all of the activities connected with harvesting are carried out by women including cutting the harvested grain, carrying the harvested sheaves, threshing, winnowing and all of the postharvest activities, though they may also be carried out by men.

Even within the same state the sexual division of labour varies enormously with activities performed by males in one district being done only by females in another. Thus, in Kanya Kumari district of Tamil Nadu, it is rare to find women harvesting, though in some parts of this district women come as migrant labourers from the adjacent Tirunelveli district during the harvest season. In part this may be a matter of caste, but historical tradition seems to also play a part. In some cases one can hazard an informed guess as to why some of these differences evolved, but in other cases even this is difficult. Unfortunately, early written materials which seem to be so rich in many respects rarely if ever say who does what. So at the moment we seem to be left in the dark as to what the practice was 200 years ago in many of these regions, and going back prior to the British period is even more difficult.

In Tamil Nadu when it comes to the sexual division of labour in harvesting, an area which includes Thanjavur district and some adjacent parts of Tirunelveli district stands apart from the other parts of the state covered by our survey (excluding Kanya Kumari where, as noted above, women hardly participate). This difference is hard to capture on any chart. In Chingleput, as in most other parts of Tamil Nadu where rice is a traditional crop, women participate in all aspects of harvesting from cutting to bundling, etc. In these areas the method of hiring is as follows: the landowner lets it be known that a given number of people are needed for harvesting the following day, and people come for the work in the early morning. Both males and females offer themselves for this work. A woman without a husband, or one whose husband was not able to work for one reason or another, is free to offer herself for work if she needs it and is physically fit (i.e. not too old or ill). She might prefer to go with relatives or friends, but that is not essential. From our observational data we find that sometimes women even out-number men in fields being harvested. In these areas the main separation among the workers is by caste, not by sex: on the whole low-caste Hindus and *Harijans* (some formerly untouchable castes) tend to be segregated from one another, either working for different landowners or in different plots. To some extent this caste distinction is being overridden today.

In Thanjavur and parts of Tirunelveli, on the other hand, harvesting is normally done by couples. A woman could not get to harvest unless she had a husband or a brother or a father to work with in the fields. As a result, widows or women whose men had other jobs or were not well could not get harvesting work. As a result some of the poorest women (those without any male earners in the household) were unable to get work during the harvest season (which is the season when workers earn close to half of their annual income). One might argue that widowers face the same problem, but this is not true since men tend to remarry unless they are too old, when they are less likely to participate in the work of harvesting anyhow, especially if they have grown-up sons.

Marvin Harriss (private discussion) has suggested that such differences must be the result of ecological or techno-environmental factors, but at this point this cannot be demonstrated. We find, for example, a large number of ecological and techno-environmental similarities between the southernmost part of Travancore and north Malabar. In the southernmost part of Travancore women do not harvest. In north Malabar they alone are in charge of all operations connected with harvesting. Both are areas with lateritic soils, with winding paddy fields separated by hills. However, the two regions have had different cultural experiences. Northern Kerala was occupied by the British for 200 years, whereas southern Travancore was ruled by a well-to-do Maharaja and the region has undergone a wide range of changes in agrarian relations. In addition, it has been subject to a major Christian influence, absent in the northern part of the state until recently. There is no data to explain how the many different forces worked to change the sexual distribution of farm tasks in this area, but from reading the literature, I am convinced that what we see in southern Travancore today is extremely different from the pattern of 200 years ago, whereas what we find in northern Kerala was less changed (at least back in the late 1950s when I first started to work there).

We do, however, have one case of a change that we can document. While the forces at work leading to this change are unique to the 1980's, it can also be seen as instructive. In two of our villages we were told that until the year prior to the time of our study, women would harvest, then after cutting the grain they would bundle it together and carry it on their heads to the courtyard or household compound of the landowner, where they would then thresh and process it. One year prior to our study, they decided to leave the sheaves in the fields so that the landlord would have to pay a man (hopefully the husband of one of the women) Rs 25 (less than a dollar at the 1991 rate) to do the carrying; if there was enough harvested, he might even have to hire two or three men. After this was done, the women would come again for threshing. In view of the scarcity of employment for males in Kerala this really is a very sensible way to get more money from landowners. In the middle of our study year, in another village south of the one just discussed, the labourers decided to adopt this new custom. Thus, during the first paddy season in our survey period, women carried the harvested sheaves, but during the second it became the work of men. If one were to look at this area thirty years from now, one might infer that the custom was the result of the load being too heavy for women despite women being seen carrying even heavier loads of firewood materials from the forest, in the same village. For those who witnessed the change, however, it was clearly a method for getting more from the landowners, and was in fact a lot more effective than a strike for higher wages would have been.

Thus, the change in the sexual division of labour in an area of Kerala which I was able to observe may be seen as a way of helping to maintain a greater income for the family in a situation where there is a decrease in the amount of work all around for both males and females and where women are struggling to preserve family income at a survival level. By refusing to carry the sheaves for free as they always did, and insisting that their males be hired to do the work, they are in part making up for the loss of work.

Impediments to female work in managing agriculture

At present, landowning women face certain constraints in managing their own land. The data from the MS project show that the most serious impediments are social, involving cultural barriers to male/female contact outside the home and outside the village. Consequently, these women seem to need help a) to buy essential inputs which require them to go to shops (normally outside the village), and to bargain with people, or even to go to the government depot and deal with the male officials there; and b) to sell their produce if the merchant does not come to their house (where merchants come to the house women are able to sell, often driving fairly hard bargains). These are male domains in which women alone do not feel welcome, unless they are accompanied by a male relative or farm servant. Apart from these two domains, many women are able to deal with the problems of supervision.

The following two excerpts from our interviews can perhaps best illustrate the above remarks.

Case 1: Selvasundaram (a widow with two sons, both employed outside the village): she owns 1.70 acres of paddy land and 50 cents of garden land. During the year prior to our study she had also taken 2 acres on lease from a relative, but in the end he cheated her, so in the year of our study she was back to cultivating her 1.70 acres. During transplanting and weeding she supervises the workers, but does not sell the produce in the market. One of her sons must do that as well as go to the market or to the cooperative society to purchase inputs. She works in the fields. If she wants any suggestions she consults her husband's brother. However, when asked about her use of fertilisers and pesticides she replied forcefully: 'I have a radio. I listen to the farm news and from that I will tell my son what sort of fertilisers we should use and what kind of pesticides to buy.' In hot seasons, she uses an umbrella and supervises the harvesting work.

Case 5: Kanaka has a husband, a son of 24 who is an SSLC failed, and another son with a BSc. Her husband has been unwell for some time. They own 2 acres of wet land and rent 1 acre in which they cultivate paddy. She supervises her cultivation daily. She does weeding along with her labourers, and also supervises transplanting and harvesting. Her husband or one of her sons shops for inputs and sells produce. She has participated in an agricultural course for women given by the Farmers Training Centre in the nearby town. She said that both her mother and mother-in-law had also participated in agricultural supervision. Her husband discusses with her and their older son what kinds of seeds, fertilisers and pesticides to use. They all listen to the radio and get information from the Agriculture Office.

What is critical in being a good farmer is the ability to attract and hold good workers, not the landowner's own physical prowess. Thus, to analyse the lower status of women (either in the landowning class or among labourers) as resulting from their lack of physical strength is to fail to understand the processes involved.

GENDER-DIFFERENTIATED WAGE LEVELS

There is today, and has been as far back as records exist, a pervasive wage differential between female and male labourers, as well as a tendency to pay less for activities done exclusively by women as opposed to those done only by men. This differential is based on the assumption that women are weaker and cannot do heavy work. Yet when the same task is done by males, it somehow becomes a hard task. We have noted in an earlier article:

one might even argue that higher wages should be paid for work which requires bending over most of the time while standing knee-deep in water, having one's legs attacked by leeches, and often not being allowed to straighten out one's back even for a few minutes by an over-zealous supervisor. Transplanting is both a hazardous and a skilled job. It is hazardous because of the illness to which it exposes the women, which include a variety of intestinal and parasitic troubles, infections, splitting heels (from standing in muddy water for hours on end) . . . and ultimately the possibility of crippling ailments like rheumatic joints and arthritis.

(Mencher and Saradamoni, 1982: A153)

I have personally seen many elderly women who, having spent years in the paddy fields, are unable to straighten their backs at all. While there is no question that manual work can affect the health of men, I have never seen any so badly bent over.

The wage levels vary considerably from village to village within the same state, even within the same district. On the whole, except where a percentage of the harvest is paid, women's work is paid less than men's. Even many of the minimum wage laws that were passed in the late 1970s and early 1980s accepted in one or another form discrimination by sex, either indirectly (as in the case of Tamil Nadu) by prescribing a lower wage for activities more commonly done by females, or directly, as in the state of Kerala (e.g. for eight hours of coolie work a woman was to be paid Rs 5.45 and a man Rs 7.88). The West Bengal minimum wage act is the same for males and females, though ploughing (exclusively a male activity) is paid at a higher rate.

The other reason often given for paying women less is that they are said to be less productive. As Agarwal has noted, in studies dealing with aspects of total farm employment, female labour time is often converted to three-quarters or half of male labour time based on an a priori assumption that female labour is less efficient. She points out that this conversion is based on differentials in wages and may not necessarily be indicative of differences in productivity (Agarwal 1985: A160–1). According to her:

Observations on comparative male-female efficiency rate, in any case, are rendered difficult when there is a sex-typing of tasks, so that there is a predominance of women in certain tasks and men in others. . . .Indeed, little systematic research has gone into actually comparing the relative efficiency of men and women in given tasks. In this overall research void, studies such as the one conducted by the Government Potato Seed Farm in Mattewara (Punjab, India) during 1973–4 to test the efficiency of different potato-digging equipment stand out. In this study, it was found that for the equipment tested, women were over three times as efficient as men. . . . This study, while by no means conclusive, is nevertheless indicative, and underlines the need for a requestioning of hitherto a priori assumptions and for more detailed research.

Agarwal 1985: A161)

In relation to paddy cultivation, no one has ever studied how much the average woman harvests versus how much a man harvests, or how many

111

seedlings a man can pull compared to a woman. Certainly there is individual variation, but there is no evidence that there is a systematic difference between the sexes. This is especially true because skill is usually as important as sheer strength in most of these tasks. In parts of Kerala where women do most of the harvesting, one never hears landowners complaining that men could do it better. Furthermore, if productivity were the sole basis for discrimination, then one might also expect differences in pay according to age or other aspects of physical well-being. The fact is that the explanations typically given are really attempts to rationalise an age-old implicit tradition of inherent inequality.

In Chingleput district, where I have been working since 1963, women always were paid less than men for harvesting. Then suddenly, in 1981, it was decided in one of the study villages that they would pay men and women the same wage (in terms of *marakkals* or local measures of paddy per day of work). The only explanation given is that 'they asked for it, so we gave'. However, it may be the indirect response to pressure by a few highly educated female landowners. One of them, in conversation, said that it was done purely on worker's demand, but I have reason to suspect that through her influence and that of others the landowners gave in without much struggle.

In paddy cultivation, females contribute a much higher percentage of the labour input than males. On the basis of a study that P.G.K. Panikar and I carried out in four villages in Kerala, Panikar has noted that more than two-thirds of the labour input is by females, ranging from 63 per cent for the second season in one of the Palghat villages to 83 per cent for the main season in the other Palghat village. He goes on to state:

> In the literature, generally one unit of female labour is treated as less than equivalent to one of male labour. In our judgement, there is no justification for this convention, especially so in paddy cultivation where female labourers attend to more strenuous and arduous work.
>
> (Panikar 1983: 77)

Disappointingly, as of 1982, the new legislation in Tamil Nadu and Kerala dealing with wage rates has maintained this wage differential (Mencher and Saradamoni 1982). The wage policy has not changed despite women officials in important government positions, perhaps because of the low representation of women in the different state legislatures. When I interviewed a female judge in Tamil Nadu, she called her subordinates so that she would have a forum to point out how unfair the law was. The men all stood there nonplussed, but I have no reason to believe that this would lead to any change.

In West Bengal, where men and women almost always do the same tasks (apart from ploughing), at least on paper the minimum wage rates are the same for male and female. However, we found this in practice in only two of our villages, one in which there has been considerable labour militancy, and the other where most of the landowners are refugees from East Bengal who had not established any traditional pattern of payment in their new village.

HARD WORK?

Despite the fact that in India ploughing is done by males, it is by no means clear that the work involved demands greater strength than other arduous work which is done *without* the assistance of animals, such as bending over transplanting for seven or eight hours a day, or carrying heavy head-loads (of grain, bricks, dirt, stones, firewood, water, etc.). One might also mention the transportation of water in large heavy pots, balanced on the head or the hip, which is done exclusively by women all over South Asia. No man would consider doing this work. Such examples make clear that we are dealing with cultural norms, not questions of physical superiority.

Dr Furer-Haimendorf, a male anthropologist, once told me that one of his informants in a different part of India had remarked, when asked why males never did transplanting or weeding work in his region: 'No man can keep standing bent over all day long in mud and rain. It is much too difficult, and our backs would hurt too much' (personal communication 1980). Yet a pregnant woman in the seventh month of pregnancy can be expected to do this work. While conversing with women, I have often heard them tell how they must ask their children to walk on their backs at night to ease the pain. In fact, tribal males do transplant in West Bengal, and we know that since partition at least Muslim males transplant in Bangladesh, but in contemporary India this is uncommon.

What is clear is that it is not physical capacity that is involved but social sanction. The ability to drive a plough or to handle agricultural implements is at least as much a matter of years of training and practice as of brute strength. Strength and stamina have to be built up, and skills need to be acquired. Furthermore, not all males are well-endowed by nature. Yet all poorer males, certainly those belonging to lower castes, must be able to plough, whereas the better-off (and better-fed) higher-caste males would consider such work below their status. In some areas, men consider the work of transplanting to be physically too difficult and it is done only by women. Agarwal argues that:

> a significant reason underlying this insistence on exclusive male control over the plough could well be to thereby establish claim over the agricultural surplus. Control over ploughing (a) means control over an operation that is usually critical for good yields (and surplus

production) under settled, intensive cultivation; and (b) provides the ideological justification for male right over that produce.

(Agarwal 1988: 536)

As stated earlier, without entering into a controversy on the question of whether or not women *should* plough, it may be reiterated that to relate the higher status of males (both culturally and in terms of power relations) with their physical strength is highly presumptuous.

Maclachlan (1983, *passim*) has used the fact that most digging in gardens is done by men as his main argument for the cultural importance given to males in the area where he worked, and by extension to all of India. Yet it is clear that at least in many parts of India, digging is considered to be extremely low-status work, to be avoided by landowning males wherever and whenever possible. As such, it is normally given to hired workers, male or female. Maclachlan has also argued that males are the natural managers because of this digging work which is more highly valued than any work done by females, and because it places them in a position to have more knowledge about agricultural operations. In the areas of our study, some male landowners (particularly those belonging to the higher castes) may know more about agricultural operations and management than females, but many females are and have been extremely knowledgeable about agriculture both today and in the past. This holds true both for females belonging to larger landowning households who do not work in the fields, and women belonging to small landowning households who both supervise and do manual labour. I would argue that Maclachlan has got the chain of causality backwards. Because men are dominant, they clearly define what they do as more difficult and more rewarding.

Regardless of energy output, when a task is performed by men alone, it is always described as *hard work*, requiring extra strength, and when it is done by women it is simply taken for granted. Thus, harvesting paddy, stacking the straw after harvesting or carrying the harvested sheaves to the household are said to be heavy tasks in areas where they have been customarily done by males, but not where they are done exclusively by females. Where the women go into the forests and carry back enormous headloads of firewood such that one can hardly see their faces, this work is not usually seen as heavy work, but in areas where such work is done by men, it is usually said to require greater strength.

DISCUSSION

It has been argued that with greater labour intensification greater yields can be obtained in paddy cultivation. Thus Burton and White state:

Labor intensification can produce a shift to male farming if the increased labor requirements exceed the time available to women after

114

allowing for their inputs to domestic production, child care, and housework. Domestic tasks take time on an almost daily basis, so the relevant constraint on women is measured in hours per day. Hence, labor intensification is most likely to affect female participation in agriculture if it requires very high labor inputs per day on a seasonal basis.

(1984: 574)

On the whole, the amount of time required for domestic chores is greater for the middle and upper classes than for the poor because when it comes to laundry a poor woman has fewer clothes to wash. In Kerala where she can bathe in a pond, she can even wash her clothes while she takes her evening bath after work. Even more obvious is the fact that a poor woman will have much less to cook as compared to the elaborate meal that the middle- or upper-class family has.

Time disposition data collected from a very small sample in one Tamil and one Bengal village and two in Kerala indicate that during the periods of the year when a landless labourer woman has no work, she spends more time on domestic chores than when she has work. In other words, during the season when there is employment available, all else is dropped so that the woman may work in the fields even on days when her husband has no work (since a man will not do women's work in the fields).

While we can readily see the differences between the traditional wheat regions of India (as well as those specialising in the lesser grains) and the rice regions under discussion here, on the micro level we may never have the complete answer as to why women in one part of Kerala do more varied kinds of work than in another. Some of this may well be rooted in history. The differences between rice on the one hand, and wheat and lesser grains on the other, relate to the political economy and sexual and caste politics in each region.

Burton and White have argued that females tend to do only those tasks in agriculture that are 'compatible with child care – tasks that are not dangerous, do not require distant travel, or are interruptible' (1984: 575). That is not what our data show. In many of the villages women migrate in search of work and commute long distances, although this is more likely to happen when there is a shortage of work in their own village; males also tend to migrate in search of work when they do not get enough work in their own villages, or when they see an opportunity to greatly increase their income by seeking work elsewhere. The other reason why men move away from a village is to seek jobs in the railways or as a motor mechanic or even work in a restaurant because of the status these jobs confer. However, some women also seek work away from home, if they are educated. When they do not, it may be either because of early marriage or restrictions on their mobility.

A trained school teacher, a health visitor or nurse may often work away from home even if she is married.

Even within the village, a lactating woman with a small infant might not travel long distances from her village. However, women are not always lactating. Furthermore, a six-months-old child on breast milk can stay without mother's milk for a longer period of time. We have found that if women get work, they tend to go for it outside their own village.

CONCLUSION

Despite all they do, Indian women are still perceived as being somehow weaker and inferior to men in many different ways. According to the rural women, a majority of husbands think of themselves as being in a superior position. There are, however, a few exceptions who were proud of their wives' knowledge and skill in agriculture. Among the poor, wife beating is common. Perhaps it is in part the result of poverty, but cultural factors are also important. Even the strongest would not dare to oppose her husband should he hit her, even if he is physically weaker or too drunk to walk a straight line. On being asked if they ever fought back with their husbands for beating them, a group of Bengali agricultural women labourers said at first that they were too weak. When it was pointed out that some of them were quite strong and a drunken man could be very weak, they all said that a wife should *never* hit her husband; they all agreed that a woman who fought back would be considered *as a bad woman*. I received similar replies to this question in Tamil Nadu.

Though the above is a digression, it indicates the power of gender-linked cultural values. The materials I have presented here deal with the sexual division of labour in rice regions and some of the related stereotypes and not with other kinds of hard manual labour done by poor females and males because space does not permit. But they too vary considerably from village to village and region to region.

In order to understand the role of gender in agriculture, it is essential to situate cases in terms of a complex set of parameters, the most critical of which are a) the nature of the crop and the type of agricultural regime followed; b) human – land ratios both in the past and today; c) caste and class and the associated questions of power and expected role behaviour; d) patterns of landownership and inheritance; e) family structure; f) cultural values; and g) the history of a given region. These factors affect gender-based power relations and cultural values, but a reciprocal relationship exists. Cultural values may also affect the agricultural regime in a given locale by putting constraints on what is possible: for example, landowners in a village without a large labour pool of low-caste women to do transplanting and weeding may choose to grow crops other than rice.

The cultural stereotypes about the role of females in agriculture have profound implications for development planning in South Asia. The view that women's work in agriculture is minimal, not as hard as that of men, at best supplemental to that of their husbands, and less productive than males is well entrenched. It leads to females being largely ignored by extension services, and when they are planned for, the planning is based on these stereotypes and often erroneous and ineffectual. Thus, there is a great deal of talk about relieving the drudgery of females, but mostly that comes down to bringing in herbicides which would eliminate the female jobs of weeding, or transplanting machines which would eliminate five out of six women from transplanting. Both of these tasks are major sources of income for poor landless and marginal households (Mencher 1985b). However, there is little emphasis on provision of water in or close to their homes to relieve them of many hours spent daily in getting water for cooking and cleaning and provide them with an expanded opportunity for vegetable gardening. These same stereotypes have contributed to the government's failure to establish meaningful reforestation and alternative cooking fuel programmes (Agarwal 1986a). These cultural stereotypes lead to poor allocation of limited resources, and to planning which can only hurt women (and indirectly their entire families) in the long run.

NOTES

1 The data presented here were collected as part of a study on women and rice cultivation (referred to herein as the MS project) in which the author has been engaged since 1979 in collaboration with Dr K. Saradamoni of the Indian Statistical Institute, New Delhi, with funding from the Smithsonian Institution, the Indian Council of Social Science Research and the Research Foundation of the City University of New York.
2 Our original plan was to study two villages from each of five districts, but later on we decided to study two villages from each of four districts, and one each from two other districts, in order to provide broader coverage and to capture differences which we had noted in certain districts of Tamil Nadu and Kerala.

Part II

DIVISION OF LABOUR IN HOUSEHOLD AND NON-HOUSEHOLD ECONOMIES

4

THE HOUSEHOLD ECONOMY AND WOMEN'S WORK IN NEPAL

Meena Acharya

Nepal
J16
J21
D13
R23

INTRODUCTION

Nepal is a country of diverse geographical landscape and great cultural diversity. Geographically the country may be divided into three east–west parallel regions: the mountains, the hilly tract and the *tarai* plains. Almost 19,000,000 people of different ethnic origins and cultural orientation inhabit this country. The overwhelming majority of them live in the hills and tarai. Despite the great cultural diversity, in terms of their attitudes towards women and their participation in the economy, the people of Nepal may be classified into Tibeto-Burman and Indo-Aryan groups. This distinction is not entirely clear cut, because some Tibeto-Burman communities have been significantly Hinduised, while Indo-Aryan communities living in the hills have been influenced considerably by Tibeto-Burman cultures. These limitations not withstanding, the variations can be usefully subsumed under these two cultural types.

The Tibeto-Burman groups mostly live in the mountains and hills. These communities do not attach great importance to protecting female sexual purity. Hence, women in these communities may move about freely and engage in various income-generating enterprises outside the household economy. Great social prestige is gained by being a good business woman, and entrepreneurship among women is priced highly. Since there is less emphasis on the virgin bride among them, marriage patterns are fairly flexible, and marriage between adults is the norm (Schuler 1981).

People of Indo-Aryan origin, on the other hand, live in the hills and the tarai. In the Indo-Aryan cultures, protection of caste and clan purity and thus control over female sexuality is traditionally of supreme importance. Descent and inheritance is reckoned patrilineally. In the more orthodox Indo-Aryan groups – for example, the Maithili community which spreads over half of the tarai area – strict female seclusion is practised. Women are under purdah and are not allowed to mix freely with people of the opposite sex, thus severely constraining their mobility. In this community

121

the only respectable role open to women is that of unpaid family worker within the household. Among the Indo-Aryan groups living in the hill areas purdah is not practised. Nevertheless, the concern with caste purity and hence with the purity of the female body manifests in various cultural restrictions, such as the mobility of pubescent girls and young women, widow remarriage, and choice of marriage partner as well as prevalence of child marriage all leading to restrictions on women's participation in market places.

However, a large proportion of the hill population from both cultural groups, particularly males, are mobile, travelling for varying periods of time to the lower regions of the country or to India for employment and trading. A majority of women in the hills, therefore, carry the main burden of the subsistence agriculture and take care of the young and the old in the household. Hence hill agriculture depends more on women than on men.

THE NATURE OF THE HOUSEHOLD ECONOMY AND WOMEN'S CONTRIBUTION[1]

Nepal's economy is predominantly agricultural which contributes more than 57 per cent to the gross domestic product (GDP) providing employment and income to more than 90 per cent of the population. According to official statistics more than 40 per cent of Nepalese do not have access to minimum food, clothing, shelter, education and health facilities. The World Bank (1990) estimates that about 70 per cent of the population nationally and about 80 per cent of the rural population have incomes below $US 150 per year, the internationally defined line of absolute poverty. The overwhelming majority of rural households, poor and not so poor, survive on subsistence agriculture. Even in urban areas the proportion of income in kind is 32 per cent of the household income. In rural areas this proportion is 61 per cent. Agricultural and non-agricultural household enterprises contribute almost 68 per cent of the total household income nationally (Table 4.1, also Table 4.2). This proportion is higher in rural areas rising to 73 per cent in the tarai. Wages and salaries contribute to only about 16 per cent of the household income in rural areas. Given the predominance of the rural sector, on an average households receive only 17 per cent of their income in the form of wages and salaries. In the case of urban households, however, this source accounts for about 32 per cent of the income. Household enterprise is the dominant source of income both for the poor and non-poor households, although wages and salaries do contribute a much larger proportion of household income for the poor.

Together with the cultural ideology of purity and pollution, the subsistence nature of the economy and mass poverty determine work patterns of women and life options in Nepal. Relating sources of household income to the various activities and allocating incomes from these sources to men and

Table 4.1 Composition of household income

	Rural				Urban[a]			
	Moun-tains	Hills	Tarai	Rural Nepal	Hills[b]	Tarai	Urban Nepal	All Nepal
Per household income in Rs	1,116	1,125	1,287	1,192	2,108	1,407	1,785	1,233
Of which %: agricultural enterprise[c]	64.3	59.6	68.3	65.2	15.5	25.3	21.4	60.8
Non-agricultural enterprises	3.7	6.7	4.9	5.4	20.0	23.0	21.1	6.8
Wages and salaries	16.0	17.3	14.2	15.6	33.0	30.1	31.7	17.3
Other income	16.0	16.4	11.6	13.8	31.5	21.6	25.8	15.1

Source: MHBS (Multi-purpose Household Budget Survey), 1984; Nepal Rastra Bank 1988.

Notes: a Mountains have no urban areas.
b Includes Kathmandu town; without Kathmandu average income is Rs 1,640.
c Includes consumption of home-produced goods including some non-agricultural items which are negligible.

Table 4.2 Composition of household income by income groups and region

	Rural			Urban[a]	
	Mountains	Hills	Tarai	Hills[b]	Tarai
Poor					
Per household income in Rs	1,002	771	764	964	807
of which %:					
1 Agricultural[c]	65.1	56.0	51.8	28.6	25.8
2 Non-agricultural enterprises	2.0	4.4	4.0	8.3	13.3
3 Wages and salaries	16.7	22.1	30.5	38.8	41.6
4 Other income	16.2	17.5	13.7	24.2	19.2
Not poor					
Per household income in Rs	1,565	1,539	1,265	2,275	1,558
of which %:					
1 Agricultural enterprises[c]	64.1	61.6	73.7	14.7	25.4
2 Non-agricultural enterprises	4.8	7.9	5.1	20.7	24.2
3 Wages and salaries	15.3	14.8	10.2	32.5	28.1
4 Other income	15.8	15.7	11.0	32.1	22.3

Source: MHBS 1984; special tabulations.

Notes: a Mountains have no urban areas.
b Includes Kathmandu town.
c Includes household consumption of home-produced goods including some non-agricultural items which are negligible.

women in proportion to their labour-input in each of these sectors, Acharya and Bennett (1981) conclude that in the rural areas of Nepal, the female contribution to total household income was slightly more than the male contribution. Women together with girls in the 10–14 age group accounted for 54 per cent of the total household income and 58 per cent of the household production. The male contribution was substantially larger only in incomes generated from work in the market sector.

The same study indicated that women's total work hours were uniformly higher than their male counterparts for all economic strata and ethnic groups. Women had to work more than ten hours per day irrespective of the household's income status. According to the time-use data from the eight case studies covered by the project on the status of women in Nepal, economic hierarchy did not make much difference to women's total work load. In fact, poor women worked slightly fewer hours than women in the middle income strata. This may be explained by unemployment among the poor women. Accordingly, in a later study (Acharya and Bennett 1983) the product correlation coefficients relating women's total work burden to both household property and household income were found to be not significant, even at the 20 per cent level. Similarly, women across all eight ethnic groups covered by the study had to work long hours per day.

Thus, despite the cultural variations between the Indo-Aryan and Tibeto-Burman groups in their attitudes towards women and women's work patterns, their substantial contribution to the rural household's survival and reproduction emerges as an overwhelming fact for both these communities. Only the content of work varied for households of various income status and ethnic origins.

The conclusions concerning the extensive involvement of women in Nepal's hill agriculture have also been confirmed by a nation-wide study of employment and work patterns (Nepal Rastra Bank 1988). Generally, women in the rural hills and mountains were found to be putting in as many hours as men in the economic sphere. This data also confirms the conclusions that in Nepal's rural areas women's total work burden is uniformly high, the minimum being about nine hours for the better-off households in the tarai. Economic status of the household did not make much difference to women's total work load, but hill and mountain women had to work longer hours than women in the tarai. Women's work load is highest in the mountains, at about eleven hours per day, slightly less in the hills, and lower in the tarai, for both poor and non-poor groups. Moreover, greater variation was observed in the kind of work women do (Tables 4.3 and 4.4). In the hills and mountains, women spend almost equal time on economic and domestic work. In the tarai, women spend relatively less time on economic work and more time on domestic work.

Definitions of work used in the above analysis are much broader than are usually understood in conventional economics, and were first used by

124

Table 4.3 Time-use pattern of women in rural Nepal (in hours)

	Conventional economic	Extended economic	Total economic	Domestic work	Total work burden
Poor					
Mountain	3.6	2.4	6.0 (6.4)	5.3	11.3
Hill	2.6	2.6	5.2 (6.0)	5.2	10.4
Tarai	1.7	1.9	3.6 (6.0)	5.9	9.5
Not poor					
Mountain	3.5	2.4	5.9 (6.6)	5.1	11.0
Hill	2.6	2.4	5.0 (5.9)	6.0	11.0
Tarai	1.5	1.7	3.2 (5.8)	5.9	9.1
Rural Nepal					
Mountain	3.6	2.4	6.0 (6.5)	5.2	11.2
Hill	2.7	2.5	5.2 (5.8)	5.4	10.6
Tarai	1.7	1.9	3.6 (6.1)	5.8	9.4

Source: MHBS 1984: special tabulations.
Note: Figures in parentheses indicate hours spent by men in similar activities.

Acharya and Bennett (1981) in their analysis of time-budget data from the eight village case studies included in the study of status of women in Nepal. In this analysis, activities have been classified into three categories instead of the usual two: economic and domestic. Participation in agriculture and industrial spheres, trade, commerce, services (for wages or income), and construction, are classified as labour force participation. This is in line with the conventional definition of labour force participation in economics. However, there is another broad group of activities which needs to be described and analysed if one is to understand properly the nature of the subsistence economy and women's roles in it. Activities such as food processing, fuel and water collection, construction and repair of one's own dwelling, and gathering of wild food, are all an integral part of the reproduction of the household economy. All these activities produce or process goods in one way or another. In the process of development they get progressively externalised and commercialised, and with their commercialisation they start being counted as economic activity. In the industrialised countries these activities are covered in labour force participation because

125

Table 4.4 Time-use pattern of women in urban Nepal (in hours)

	Conventional economic	Extended economic	Total economic	Domestic work	Total work burden
Poor					
Hill	2.7	1.9	4.6 (6.0)	5.3	9.9
Tarai	1.3	1.4	2.7 (6.8)	6.7	9.4
Not Poor					
Hill	1.8	1.2	3.0 (5.5)	5.6	8.6
Tarai	1.3	1.0	2.3 (6.1)	6.6	8.9
Urban Nepal					
Hill	1.9	1.3	3.2 (5.5)	5.6	8.8
Tarai	1.3	1.1	2.4 (6.2)	6.5	8.9

Source: MHBS 1984: special tabulations.

Note: Figures in parentheses indicate hours spent by men in similar activities.

they are performed commercially. In the above analysis this group of activities has been considered part of the subsistence economy.

Finally, cooking, servicing, cleaning of the house, dishes and pots, doing the laundry, shopping, and other domestic chores such as childcare have been classified under domestic activities. This group is the core of the household's reproductive process and currently not amenable to economic measurement.

WOMEN'S WORK PATTERNS

Returning to the issue of women's work patterns, landownership and the status of household property, degree of strictness in the practice of female seclusion, topography of the area, accessibility or inaccessibility of women to markets and market skills as well as demographic factors such as the presence of male adults or adolescents and young children in the household determine the kind of work women choose to do or not to do. In the hill regions of Nepal, women across all economic strata perform agricultural field work. Women in the lower economic strata participate in wage markets as well. Women in the tarai, however, resist field work even in their own fields as long as their participation is not absolutely required. However, a large proportion of grain-processing activities takes place within the family courtyard, and women actively participate in this work – in fact they dominate it. In addition, food for workers in the fields, cooked in the house, keeps women occupied the whole day during the busy agricultural seasons.

Even in the tarai women of the lower economic strata were found participating extensively in seed selection, preparation and application of manure, planting, and harvesting of agricultural crops (Acharya 1981). The strenuous task of hauling water and fuel is, of course, largely women's work all over Nepal. So is the work related to the care of large and small animals all over the tarai and hill areas except where herding is the major activity pursued by males. In these communities men herd animals for extended periods of time, and perform most of the activities related to their care (Molnar 1981). In such cases agriculture is virtually controlled and run by females. Women form the majority of the household workforce engaged in agriculture, and their labour input in major crops is slightly more than that of men (Acharya 1984). Women were extensively involved in planting, harvesting and post-harvest processing of these crops, as well as in the seed selection and operations of fertiliser application. In some areas crops like maize or millet and other minor crops are dependent exclusively on labour input by females (Bennett 1981).

In the hill areas of Nepal and in the households of small farmers in the plains, women participate actively in decisions pertaining to farms as well. In the aggregate data of the eight villages covered in project on the *Status of Women*, women were found to have made 43 per cent of these decisions by themselves. They were involved with men in another 12 per cent of such decisions taken in the household. Management decisions considered in this context included issues of labour allocation, crop choice and allocation of fertilisers. Despite the generally lower status of women in the Indo-Aryan groups as compared to women of the Tibeto-Burman groups, female input in decisions on farm management was uniformly high for all hill villages in the sample except in a remote Tibeto-Burman village where women were engaged more in trading than in agriculture.

Thus, the variations in the kind of activity women do depends on many factors, among which cultural ideology and economic necessity emerge as the major ones. Acharya and Bennett (1983) use a four sphere model to examine the role of various factors in determining patterns of women's work (Figure 4.1). The village economy is conceived as operating in four concentric spheres, each of which offers a set of possible strategies for increasing the family's living standard of welfare by the generation of income, production of goods, performance of services or generation of leisure for each individual member. At the centre of this model is the *Household maintenance* sphere, analytically separated from the other economic or income-earning sectors, but supporting all of them with its services. Sphere I incorporates 'domestic activities' as defined in the previous section.

Sphere II of the rural economy is the *Family farm enterprises*. This is essentially the household as a unit of production and is to a large extent

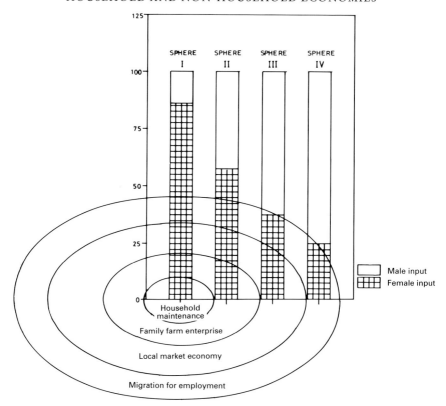

Figure 4.1 Male/female input into the four spheres of the village economy

Source: Acharya and Bennett (1983).

anterior to the market economy. Although not primarily destined for the market, the activities within the sphere produce goods distinct from the services which are included in sphere I. These goods can be assigned an economic value with a fairly high degree of accuracy if the data collected on household production is extensive enough. Agriculture and animal husbandry from the labour force participation group, and subsistence economic activities, as defined above, are included in sphere II. The delineating feature for sphere II is the fact that all these activities take place within the limits of the household economy, i.e., in their own farms and households.

Sphere III in this model is the participation in the *Local market economy* which includes work performed in the village or nearby bazaars for wages and profit either in kind or cash as well as work performed at home to produce goods primarily for sale. Thus, apart from wage work and trading, manufacturing primarily for sale is included in sphere III. Since most women who were engaged in manufacturing for sale were also trading the goods

128

themselves, it was not possible to separate 'manufacturing' at home from trading in the local market meaningfully.

Sphere IV in this model refers to short-term migration for employment and participation in the *wider market economy* beyond the village. The distinguishing feature between sphere III and IV, once again, is the *locus* of the activity. Any employment or work including portering, army service, trade or agricultural labour that required a household member to spend the night outside the village was considered as participation in sphere IV.

In proposing such a model of the rural economy, the authors are quite aware that the boundaries among the adjoining spheres are fluid, and that 'most work' activities belong somewhere on a continuum moving from domestic through subsistence to the market rather than to discreet sectors. Criteria used for the classification of activities into one or the other of the four spheres have been multidimensional. The physical locus of work, the social unit in which it takes place, the type of output (i.e. product or services) and the destination of the output (home consumption or sale) – all have been used as a basis for including a particular activity in one or other group. The boundaries can be defined differently depending on which criterion one wishes to focus on. However, for this analysis, the most important feature of these spheres is that women participate to very different degrees in each of them.

Women were responsible for 86 per cent of the household time input in sphere I and for 57 per cent in sphere II. In the market sector, i.e. in spheres III and IV, time input by females is substantially lower at 38 and 25 per cent respectively. Thus, men in Nepal as elsewhere are much more likely than women to combine their work on the family farm enterprise with work in the market economy. They can do this not only because they have a comparative advantage in marketable skills and capital endowment, but also because female labour is available within the family to continue the production process in the family farm enterprise and to take care of the old and the young. This dependence of male availability of market activities on the subsistence agriculture exclusively supported by women has been extensively discussed in the Latin American context (Aguiar 1982). In the Nepalese context Blakie *et al*. (1980) have highlighted this aspect of Nepal's hill agriculture.

However, there are significant inter-community variations in the degree to which women participate in the market sector, i.e. in the third and fourth spheres (Figure 4.2). The pattern displayed by the Indo-Aryan group limited women to the second and third spheres to a large extent. The Tibeto-Burman group, on the other hand, involved women to a significantly greater extent in the market sectors.[2] In the Indo-Aryan communities, the women's contribution to sphere III was 30 per cent at the most. Comparable figures for the Tibeto-Burman group ranged between 40 and 68 per cent. Similarly, women's contribution to activities in sphere IV was relatively insignificant

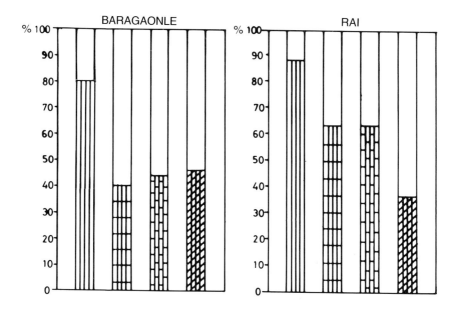

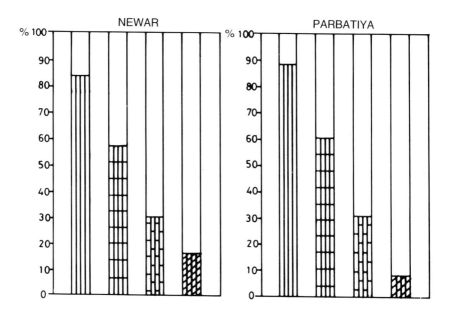

Figure 4.2 Relative male/female input in the rural economy of villages
Source: Acharya and Bennett (1983).

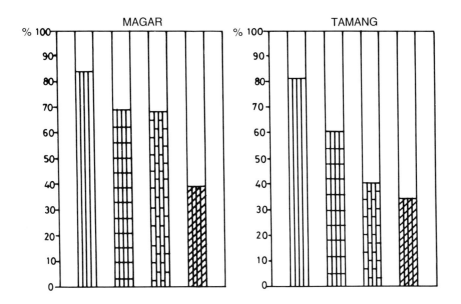

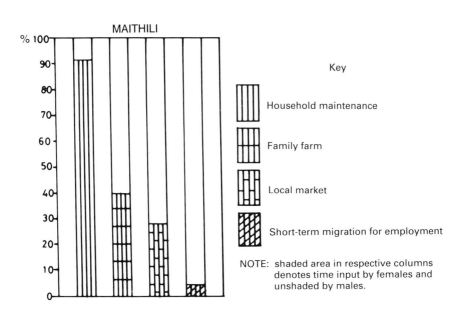

Key

Household maintenance

Family farm

Local market

Short-term migration for employment

NOTE: shaded area in respective columns
denotes time input by females and
unshaded by males.

131

(less than 15 per cent) for the Indo-Aryan groups. On the other hand, the Tibeto-Burman women were found to account for 34 to 46 per cent of the total household time spent in this sphere.

On the basis of this descriptive model Acharya and Bennett (1983) ran some regression models relating women's participation in these four spheres to some behavioural variables known to be distinct for the two cultural groups. These included age of marriage, distance to natal home, social perceptions about women's character, female mobility and freedom in the choice of marriage partners. In addition, the household income and asset status and demographic variables such as ratio of young and adolescent children in the household, the number of adult men and women in the household, the age of the individuals, and family structure (whether nuclear or extended) were also entered in the equations as independent variables. In the Tibeto-Burman groups girls were observed to be married at a later age than among their Indo-Aryan counterparts. Similarly, Tibeto-Burman women had greater mobility, perceived a more positive image of the female gender, could choose their own marriage partners as opposed to the practice of forced child marriages in the Indo-Aryan communities, and they tended to marry nearer home than their Indo-Aryan sisters.

As expected, most of the cultural variables associated with the Tibeto-Burman groups were found to have a positive effect on the participation and labour supply by women in the third and fourth spheres. In this group we may cite age at first marriage and female mobility. On the other hand, distance to natal home had a negative effect on these activities, as expected. Social perception about the female image turned out to be a positive sign in the sphere IV equation. But, somewhat puzzlingly, in the sphere III this equation had a negative sign. Furthermore, asset or income variables were found to have no effect on women's work patterns. Among the demographic variables, women's age had a positive association whereas the number of adult males in the household had a negative effect on decisions regarding labour supply. The number of girls in the 10 to 14 age group in the household, however, had a negative effect on women's market participation in both the spheres, which ran counter to expectations.

I have extended the discussion a step further and analysed the women's labour participation in the market and number of hours worked (hours supply) separately (Acharya 1987). The hypothesis is that participation and decisions of hours supply are guided by different sets of factors. Accordingly, probit model is used to test the probability of male–female participation in the labour market. As a second step, Least Square Estimations are used to estimate the supply of work hours conditional on participation. In this analysis all market participation, i.e. sphere III and IV in the above discussions, have been lumped together (See Acharya 1987).

The result of these estimations in general support the conclusions derived earlier that cultural ideology plays a major role in women's decisions concerning participation in the labour market. Some of the inconsistencies observed in the earlier regression analysis could be more easily explained now.

Among the caste and ethnicity variables, caste was found to have a negative effect on decisions pertaining to labour participation for both men and women. The higher the caste, the lower is the probability of participation. This was expected, since both men and women of higher caste generally do not like to work for others, particularly if the work involves physical labour. But caste had a positive effect on hours spent by women. Moreover, the elasticities of hours spent with respect to caste were found to be quite high. Similarly, most of the cultural variables positively associated with Tibeto-Burman groups had a positive effect on women's participation in the market, but negative effect on their labour supply.

It seems that while there is a greater probability that a Tibeto-Burman woman will participate in the market than will her Indo-Aryan sister, once in the labour market the former will work fewer hours than the latter. These findings may be attributed to regional differences in market opportunities rather than supply factors. Once in the labour market, women from Indo-Aryan groups and/or higher castes have greater work opportunities. This is because, in this data set, three sampled Indo-Aryan groups belonged to geographical areas either near Kathmandu or in the tarai, and had better opportunities of finding work in the market. The Tibeto-Burman group in the sample, on the other hand, were from interior areas in the hills and mountains and had fewer chances of finding work in the market.

Most of the demographic variables were found not to have a significant effect on the decisions regarding labour participation. The number of children under the age of 9 in the household had a negative effect on the probability of women's participation in the market, but their effect on hours supply was positive. In Nepal, women in rural areas can take young children to work. Moreover, children between 5 and 9, especially girls, have been found to perform various household duties, thus relieving women to work in the fields. Children between 10 and 14 years of age had a positive effect on decisions taken by women to participate in work but a negative effect on hours supply. Children in this age group seem to share market work with women which explains their positive effect on market participation. But once in the market they substitute for women, hence the negative effect on hours supply. This is as anticipated if one assumes that hours worked are basically determined by demand factors. The number of male adults in the household did decrease the probability of female participation in the market and hours supply of women. The effect of the number of female adult members in the household had a positive effect on both of these variables. The larger the number of women in the household, the greater is the probability of her participation in the market and the longer the hours of such work.

The size of agricultural assets, as expected, had a negative effect on the supply of market labour as well as women's decisions to participate in the labour market. However, the elasticities of female labour supply to the market was found to be positively related to the amount of agricultural assets in the household against the proposed hypothesis of negative relationship. This could be explained by estimation problems. Alternatively, they could reflect the fact that women coming from landed households have better opportunities for salaried employment, once they are willing to work in the market. Further, since the participation by women in the Tibeto-Burman households in the market place is not stigmatised as is the case with the women in the Indo-Aryan groups, the former group generally participate in trading irrespective of their household's economic status. On the other hand, more assets provide greater capital for trading and hence more opportunities for market participation.

As expected, wage rates had a positive effect on both participation and hours supply in general.

SUMMARY AND CONCLUSIONS

The Nepalese economy is predominantly agricultural, and most of the production takes place within the household. Most of what is produced is consumed at home. An overwhelming majority of the people are poor, and women in these households cannot afford not to be engaged in productive activities of one kind or another. Therefore, women from all economic strata and all three geographical regions of Nepal are found to be working at least nine hours per day. The kind of work they do, however, does differ according to the economic status of the household, caste, ethnic origin and geographic region. Despite the variations, overall women and girls between the ages of 10 and 14 were found to be contributing 54 per cent of the household income.

Besides working on their own family farms, the rural poor women of Nepal were found to be spending a significant proportion of their time in wage work also. Landownership and income status of the household tended to reduce the wage work women did, but did not affect the total work load they bore. Poor women in the tarai, on the other hand, performed less of the economic and extended economic activities, probably because they could not get enough socially acceptable work in the market and lacked the capital and skills to generate income by working at home. However, this could also be a problem of recall, since the Nepal Rastra Bank data is collected by the recall method. Even in the Maithili tarai village covered by the project on the status of women in Nepal in which data were collected through observation, women from the lower economic strata were found to have spent more time in economic and extended economic activities.

The cultural ideology of purdah and caste taboos did play a decisive role in the choice of work for both men and women, but more so for women. The higher the caste of a woman, the lower is her participation in market activities. Ethnic groups which attached lesser importance to the purity of the female body accorded greater freedom to women in the choice of work. Hence in Nepal, women from the Tibeto-Burman group, less concerned about this, were found to be participating in market work to a much greater degree than women from the Indo-Aryan group, where the ideology of clan purity and control over women's sexuality is carried to obsessive limits.

Work patterns and life options of women are very much controlled by caste and cultural taboos as to what one can and cannot do. Some women break out of this simply because dire economic necessity compels them to, while others try to cross this barrier by acquiring education and new skills which will help them obtain new kinds of jobs. Still, in the Indo-Aryan groups the need to control female sexuality is so strong that the taboos stick even in the new kinds of jobs. For example, a Brahmin woman would hardly like to work in a shoe factory, because shoe-making is so negatively stigmatised (since it involves the extremely polluting work of processing the hides of dead animals). The concept of women being the 'bearers of family honour' is so ingrained that without the family's blessing she cannot perform any work other than what is traditionally allowed (Lateef 1986). Kemp (1986) describes how poor women in a southern Indian village had to choose between hunger or social humiliation. Households faced this dilemma since sending their women to work in the market involved their social degradation. Epstein (1986) describes how Muslim women were found going back to purdah in Pakistan as the family's economic status improved. This ideology of female seclusion has led to greater impoverishment of poor households in the tarai than in the hills. This is because poor women there have much less choice in the kind of work they can accept. As a consequence, they experience a greater number of unemployed days, resulting in poorer households. This is true of men as well, but to a lesser degree.

Decisions about labour supply have two components, first, decision to participate in the labour market, and second, about how many hours to work. The property, caste status and ethnic factors seem to operate more on participation decisions than on hours to be spent. The latter was found to depend more on demand factors and some demographic factors on the supply side. Thus, while the caste and ethnic characteristics associated with the Tibeto-Burman group feature prominently as positive factors in the probit analysis, some of them emerged with contrary signs in the equations related to hours spent.

Finally, the framework of analysis presented above is also interesting from a methodological point of view, since women's studies are still in an evolutionary stage. The analytical framework used in Nepal has integrated both qualitative and quantitative traditions of data collection and analysis. It has also tried to use the quantitative analytical framework available in traditional labour economics. The Nepal group would be very happy if it can make a contribution towards evolving an effective analytical framework for studying women's work patterns and decisions of market participation in the region.

NOTES

1 Most of the findings discussed here are based on four primary sources. They are Acharya and Bennett (1981, 1983); Acharya (1987); and Nepal Rastra Bank's multipurpose household budget survey data set.
2 In the analysis, Parbatiya, Maithili and Newar are classified as Indo-Aryan whereas Baragaonle, Rai, Megar and Tamang are included in the Tibeto-Aryan group.

5

THE HOUSEHOLD AND EXTRAHOUSEHOLD WORK OF RURAL WOMEN IN A CHANGING RESOURCE ENVIRONMENT IN MADHYA PRADESH, INDIA

Deipica Bagchi

Nearly 300 researchers from Asian and other countries participated in an Asian Regional Conference on 'Women and the Household' in New Delhi, India in the winter of 1985. The significance of the conference was in its wide coverage of issues and participation from the developing world. One of the conference's sub-themes on women and home-based production generated a consensus highly relevant to the content of this chapter, i.e. the invisibility of home-based work which is tied to a general neglect of the household economy and a narrow definition of work which precludes its inclusion in official statistics. A precondition for correction of the problem, the participants concluded, is to make this work visible through research and publications in order to help mobilise social and legislative action.

This chapter is an attempt towards this larger goal. Some specific suggestions, framed as conference recommendations, are dealt with in the concluding section of this chapter.

Rural systems in their unique socioeconomic and ecological inter-relationships have always been of interest to geographers. Set in specific time–space context, rural societies convey specific messages of unique man–land relationships. However, the interest in social relations of production, particularly of the gender aspect, is a recent development following the publication of Ester Boserup's work on women in economic development (Boserup 1970). The author's revelations of widespread negative impacts of development on women eventually led to the declaration by the United Nations of the International Women's Year (1975) and the International Women's Decade (1975–85). Subsequently, in almost all disciplinary fields in the social sciences interest has surged in the revaluation of societal processes in the context of gender.

This chapter is a product of the last ten year's research on women's work in rural Madhya Pradesh, India. The study is on understanding of gender in sustaining households and production lines. The focus is on women's work roles in a space–time context in the domestic and the extradomestic sphere. Specifically, the study maintains focus on two separate and distinct spheres of women's work within the household that are socioeconomic in nature, none the less being persistently invisible to society. The essential household role is nurturing in nature aimed at providing a 'good home life', while catering to four basic needs i.e. food, clothing, shelter and overall care for the young and the old. A smooth-functioning household not only ensures reproduction, but also aids in freeing its members for work in the market economy. Household work of this nature falls into the category of 'necessity'.

The other area of household work, termed in this chapter as 'extrahousehold' pertains to what Papanek describes as 'Status Production' work, enabling families to 'utilise women's work in ways that are deemed more productive for the entire family' (Papanek 1989: 98). Women's activities, in Papanek's view, may range from unremunerated support services which otherwise could only be performed through hired labour, work that is geared towards delayed rewards for the family, to activities that directly affect the status of the household in relation to others in the community or in the 'patronage network'. Some of these may not be so clearly definable as those producing economic rewards. But, all of such activities are for what the household values as above 'necessity' (Brown 1982: 152–3). Households utilising women's work do so to enhance productivity of the entire family. Even withdrawal of women from paid work might turn out to be a strategy for further mobility rather than an end product of mobility. The truth is that women engaged in 'status production work' sacrifice their own autonomy in the interest of collective benefit.

Obviously, the categories of work put together are crucial; for most they are a source of sustenance and survival, while for some others they might increase the competitive advantage of the households in owning and controlling assets other than their own labour. Both categories of activities, however, lack clear definitions of purpose apart from being unremunerative – two plausible reasons for their invisibility as well as exclusion from established economic measurements. The illustration, however, Figure 5.1 symbolising the domestic economy, includes only those activities that are the focus of this study.

A conceptual model of the household work sphere is presented in Figure 5.1. The title 'Rural women's work "Chakra" ' has several symbolic connotations. 'Chakra' is a Sanskrit word for a circular form possessing an intrinsic property of movement, popularly implying (the form of) the wheel. The household economy is viewed as symbolising one of the wheels of society, responding and adjusting to the other wheel which is the market economy.

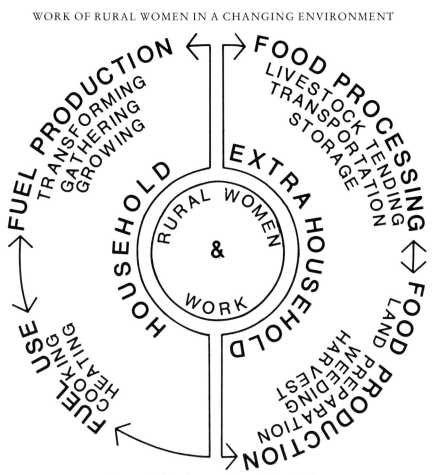

Figure 5.1 Rural women's work 'Chakra'

The pre-industrial economies were fully cognizant of the need for balance between these wheels that sustained the society while propelling it towards growth. The post-industrial devaluation of the household in favour of the market economy has occurred at the risk of disturbing this balance between the two wheels. Market incentives have inevitably led to continued growth of the sector, usurping in the process the demand for the goods and services offered by the household sector. Labouring under the stress of this loss, the household economy has been further weakened by being branded as merely a consumer group, eventually forcing members to enter the labour market as wage earners. The extrahousehold work of women is an attempt, in whatever manner, to link with that other economy.

The circular form of the 'Chakra' also represents a continuum – work without beginning or end. While services offered by the market economy may be availed at specific blocks of time, the household services must be made available in a personalised manner at any hour of the day dependent on the needs of individual members who might be at different stages of the life cycle. Additionally, certain specific household services, such as serving meals to children and the elderly which must be provided at specific hours, create a rigidity of schedules that are part of many limitations facing women in choice of work. More importantly, the nature of goods and services offered in this sector are non-substitutable by the market economy.

Thus, women are albeit positioned at the centre of the 'Chakra', signifying their central role in running of the household economy. Yet, one must also not lose sight of the negative symbolic connotation of this position – of being trapped by tradition and culture within the confines of the household space.

The household work sphere must also expand whenever needed to include extrahousehold work that is income substituting, earlier referred to as 'status production work'. Considerable overlap might exist between the household and the extrahousehold work, the former expanding to use up extrahousehold work time during special or crisis occasions, and the latter expanding to encroach upon household work time during peak agricultural periods. Women learn to juggle the two spheres, balancing the demands of both sectors.

The research was conducted at two separate periods – in 1979 to 1980 and in 1983 to 1984. The field work was spread over diverse ecological regions of the state in order to capture role variations of women by region and culture. The state of Madhya Pradesh possesses a composite of ecozones ranging from tropical humid to semi-arid, and distinct crop regions of wheat in the north, rice in the east, cotton in the west and millets in the south presenting an ideal laboratory for analysing women's work in diverse economic settings. Further, it's location between the classical north–south culture zones has made it a cultural shatter belt subject to cultural-historical influences from the Aryan north and the Dravidian south professing variant views on women's status in society (see Raju 1987).

The selection of districts was thus to capture this cultural diversity within the state (Figure 5.2). The district of Morena in the north represents the wheat-corn belt and influences of northern culture; the district of Raipur in the east represents the rice economy and culture; the districts of Khandwa-Khargone represent the cotton economy and the Deccan culture of the south. Additionally, the districts of Betul, Dhar and Dewas, in the southcentral and western parts of the state were selected to represent the fuel crisis situation in the forested as well as the semi-arid parts of the state.

Selection of random sample points (villages) within the chosen districts were made to ensure true representation of the geographical or cultural diversities within each district. Roughly, 10 to 15 per cent of the village households were randomly sampled. Care was taken to stratify the village samples by tribal–non-tribal population and by economic categories. The

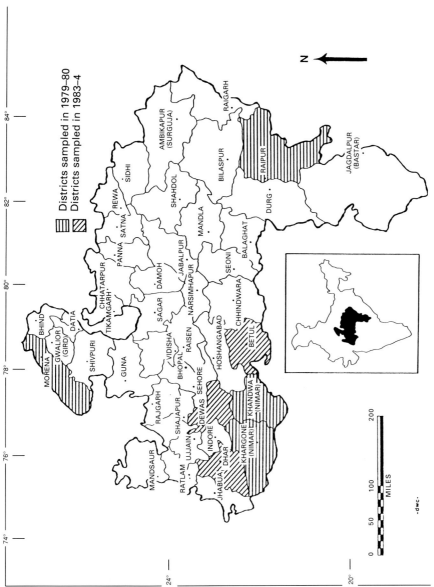

Districts sampled in 1979–80
Districts sampled in 1983–4

Figure 5.2 Madhya Pradesh

Table 5.1 Household and extrahousehold work

	Cases reporting[a]	% of total[b]
Household work		
Cooking	63	92.8
Collection of water	61	88.4
Collection of firewood	56	81.2
Making cowdung cakes	56	74.0
Collecting cowdung	50	72.5
Collecting fodder	30	45.0
Grazing livestock	30	45.0
Extrahousehold work		
Wage work	44	63.8
Crop processing	38	55.1
Marketing	28	39.7
Work on own farm	26	38.2
Craft	13	18.8
Voluntary work	7	10.3
Other[c]	6	8.7

Source: Survey data by author, 1983–4.
Notes: a Roughly 75 per cent of the sample constitutes respondents from low income landless population.
   b Total cases surveyed (N=69)
   c Activities reported under 'other' included construction, dairy, vegetable, gardening, etc.

respondents were interrogated through a structured questionnaire, with utmost care during the interrogation to record responses while eliminating bias of either the interrogator or the respondent. Female respondents were interviewed often accompanied by, or in the company of, their spouses.

Energy and food systems are two major and interacting elements in the rural economic systems. Sufficient food, enough fuel to cook and an environment capable of providing both are essential to a healthy society. Further, energy and environment are critical to rural women for meeting their family's basic needs of food and shelter. An unbroken link exists from the field, farmgate to the domestic hearth that connects the household and the extrahousehold work spheres of rural women (Table 5.1).

In Figure 5.3 these activities are ranked by the frequency of cases in the sample reporting different types of work. The list of work is by no means exhaustive as only those activities which are economic in nature or directly impinge on the household energy situation appear in the list. Cooking and collection of fuel and water are a major responsibility for most. Farm-related work on family farms, wage work on farms, crop processing and livestock care appear next in the ranking. The least values are for craft or other work of a subsidiary nature.

Several of these household and the extrahousehold activities stretch

142

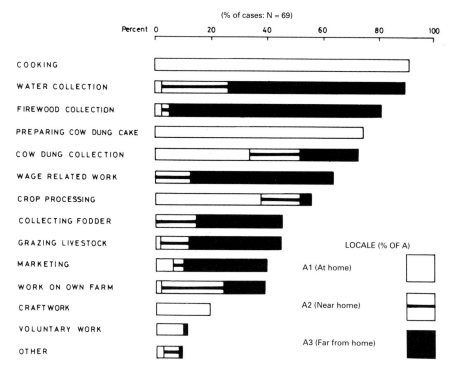

Figure 5.3 Rank order of activities performed by women

Source: Survey data by author.

Note: Activities reported under 'OTHER' are construction, dairy, marketing of vegetables.

women's work space farther from home involving an inordinate amount of travel time. Women's daily hours are frequently fragmented into a multiplicity of overlapping and interrupted tasks. When to the given list are added the essential service-related tasks of washing, cleaning, feeding, tending of children and the elderly, and neighbourly duties, women are found constantly juggling time and space to accommodate the 'economic' tasks with their 'service' roles, the 'invisible' or the 'undervalued', work.

FUEL COLLECTION – A MAJOR HOUSEHOLD CHORE

The rural household is now recognised as the major consumer of woodfuel and is at the centre of the rural energy crisis (Agarwal 1986a; Bagchi 1987; ILO 1987). The principal sources of energy in rural systems are human, animal power, and biomass in various forms – wood, crop residues, animal waste – used in the two major areas of the economy, agriculture and the domestic sector. The energy and the environmental crisis that has become synonymous with deforestation and desertification is the product of pro-longed overuse of this freely available biomass resource.

Women's work, more than men's, depends on access to energy and biomass. The most significant use of the biomass is in cooking. The same fuel is also used for water and space heating, and also in processing industries at the household level which are important income- and employment-generating sources. Rural women are caught in the midst of this crisis, working already long hours to produce enough food and income for the family's survival, and now spending longer hours in collection and process-ing of fuel in a situation of growing household needs.

How far really has the fuel scarcity affected women's work lives? An attempt to chart out the precise inter-relationship produced a system's diagram presented in Figure 5.4. Fuelwood has been the most essential raw material for women's most important and time-consuming activity of food preparation – a single activity that combines multiple tasks such as collection and transportation (headloading) of water and fuel, processing of food, feeding, and cleaning up. Five to eight hours each day may easily be spent on this chore alone (ILO 1987).

If wood is scarce, additional time must be spent in collecting and patting cowdung into fuel cakes. Traditionally, collection of cowdung was not time consuming because of its adequate supply in the village whereby children could be involved in collection. However, this resource is fast becoming scarce in the wake of its new technological uses in biogas production. Alternatively, bushes, twigs, roots, crop residues are increasingly being used which must be collected from fields and woods, near and afar to keep the home hearth burning. In desperate situations, limbs from live trees are cut and burnt accentuating the ecological crisis. This has a negative impact on women's health by way of physical discomfort of smoke and longer cooking hours. The alternative biomass products used by women for fuel – branches, leaves, straw, dung – impinge critically on their other important uses: crop residues for fodder, bamboo and branches for housing, sheds, baskets, dung for wall–floor plaster, and for fertiliser. Thus, a heavy trade-off is involved.

Table 5.2 presents a profile of activities related to fuel collection. Ninety per cent of the sample population traverse long to longer distances for collection; for 60 per cent, the activity consumes from half to a full work-day; for one-half of the population the activity is a daily routine; around one-third considers the time spent in collection a major problem; only 11.5 per cent of the population encounters no problem in fuel collection.

The collection of cowdung – the alternative to wood – is an easier chore because it is mostly collected in the village and vicinity. Young girls are often entrusted with the task of collection. Dependency on cowdung is primarily in the lower-income groups who find themselves in a dilemma as the available supply is funnelled into the new biogas technology being intro-duced as alternative energy. Ironically, the biogas energy is beyond reach for

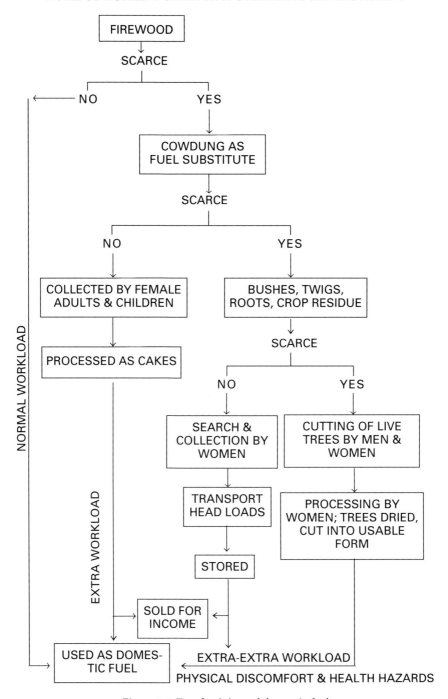

Figure 5.4 Eco-feminism of domestic fuel

Table 5.2 Firewood collection

	Total Cases	%
A. *Distance covered in collection*		
Total	61	100.0
Very far from home	50	82.0
Far from home	05	08.2
Near home	03	04.9
No response	03	04.9
B. *Hours spent in collection*		
Total	61	100.0
Whole day	22	36.0
> 4 hours	14	23.0
2 to 4 hours	16	26.2
1 to 2 hours	05	08.2
No response	04	06.6
C. *Frequency of collection*		
Total	61	100.0
Daily	31	50.8
Weekly	26	42.6
Monthly	01	01.6
No response	03	04.9
D. *Specific problems cited*		
Total	61	100.0
Excessive time required	19	31.0
Restrictions by forestry	03	04.9
Both of above	21	34.4
Other	07	11.5
Encounter no problems	07	11.5
No response	04	06.6

Source: Survey data by author, 1983–4.

precisely this population which is desperately looking for alternative fuel sources. The new plan, like many of the rural development plans, is likely to provide little benefit to those most burdened with the fuel crisis.

The fuel issue affects, in a variety of ways, the area of women's most time-consuming household work: cooking, which alone consumes, on an average, four to five hours daily. It takes even longer with inferior fuel. The traditional mud-stove, widely in use and suited only to burning of wood, produces much lower energy efficiencies if used with inferior biomass. The new models of energy-efficient stoves are yet to become popular especially among the energy-affected group due to their economic marginality. The diffusion of energy-efficient stoves is similar to that of the biogas technology

– restricted only to a few large landholders in each of the villages surveyed. Had the fuel development programme been realistically devised, the community-owned biogas as opposed to individual ownership should have been the focus.

Another viable option for future fuel supplies pertains to wood-fuel plantations which are state run with financial aid from the United States Agency for International Development (USAID). The Forestry Department now has a wing for Social Forestry responsible for managing forest plantations on waste lands. The community's involvement and participation in the project was found to be very weak, particularly by the women's group whereas empirical evidences around the world now document communities and women's participation as key elements in the success of such afforestation plans. Women stand to gain significantly from social forestry since they are the ones hardest hit by deforestation.

From the sphere of the household, let me now move on to the most important extrahousehold role that women play as active participants in the food-production system. Repeated research has revealed women to be the key link in the energy–water–food chain. Women are responsible for producing more than half of the food produced in the developing world, definitely more than 60 per cent of the food produced in the subsistence sector (Sevard 1985). Conducted outside the formal market economy, women's work in this sector has remained invisible. Documentation of this sector was hopelessly neglected until the International Women's Decade.

The informality of the work environment in this sector imposes the burden of irregular wages and work hours, easy retrenchment in periods of crisis and no guarantees on security of employment or benefits. Yet this sector is crucial to the economy and helps in the survival of a large population group.

The following section presents an empirical evaluation of women's work in a changing agricultural environment in Madhya Pradesh, India. The agricultural system here is of traditional subsistence type bearing imprints of the feudalistic and colonial modes of production. The traditional division of labour has absorbed shifts, off and on, towards greater or less female participation dependent on the need or lack of it for production intensification. Current influences of modern technology are producing further readjustments in the existing division of labour. A closer inspection of the gender segregation of labour reveals fewer tasks reserved exclusively for males. Women are abundant in laborious farm-related tasks, shouldered in addition to their customary 'nurturant-type' activities relating to supervision and care of the family.

The study is conducted in three major crop regions of the state – the wheat-growing region of the north, the rice region of the east, and the cotton-growing region of the south west – each having benefited from the Green Revolution. The sample is also designed to reflect the north–south cultural polarity in India in terms of social attitudes towards women. The northern region falls into the global context of what Boserup calls, 'the West Asian (or the North African-Arab) pattern' of low female profile in agricultural operations imposed by the profound cultural influences of Islam. The southern tracts of the study region resemble Boserup's South Asian pattern of high female profiles in farming (Boserup 1970: 15–31). Also, the farming system is largely dependent on the use of the plough that is presumed to restrict female participation except in the intensively cultivated regions where high demand for labour necessitates dependence on the female labour force. According to Raju, in a conducive social atmosphere higher labour input by females in rice-growing areas reflects an enhanced response to labour market conditions. She further maintains that, other things being equal, region-specific cultural or societal norms remain vital in defining female roles in agriculture (Raju 1987).

The regional samples also include proportionate representation of the tribal population for comparison of female work patterns in tribal and non-tribal population groups. It was hypothesised based on the prevailing contention that women of tribal communities enjoy a greater degree of autonomy than their non-tribal counterparts, and are likely to be more active in farming. The study results reveal the two groups much less variant than was hypothesised, confirming a good degree of assimilation of the tribal sub-culture in the prevailing social ethos, a process of Sanskritisation as identified by sociologists (Srinivas 1956; Kalia 1961).

The farm activities are grouped in three stages (Figure 5.5). The initial stage of field preparation involves ploughing, levelling and fencing of land. The next stage is of sowing, including broadcast, weeding and watering. The final set of activities are of harvest, binding of sheaves and storage.

The data confirm Boserup's thesis regarding highly restrictive female roles in the operations of the plough. In the successive operations of land levelling and fence building (bunding) women work jointly with men to a much larger degree in the paddy region compared to the wheat region. In the cotton region women demonstrate a higher level of participation than in the wheat region in building and mending of fences. Nevertheless, this phase of operations is primarily carried out by males.

Women's farm work compounds during the next stage of the crop cycle – that of sowing and weeding. In sowing operations, highest female participation is in the cotton district – a large majority of women managing it alone – followed by the wheat region where three-quarters of the sample reports joint male–female roles. Surprisingly, least female participation is demonstrated in the rice region.

Paddy is generally sown in India the transplanted way – seeds sown in nurseries to be transplanted later in the prepared fields. The work is highly labour intensive, heavily dependent on female labour. Additionally, the sample of the paddy district carries a high proportion of the tribal population which should have been reflected in high levels of female participation in the sowing operations. The anomaly in the data of lowest levels of female work input in sowing, almost wholly in conjunction with male labour is because of the existing method of sowing paddy by broadcasting in the sample district that eliminates the need for female labour.

The successive weeding operation of *Bayasi* in tribal regions also excludes female labour because of the use of the plough in weeding for loosening and removal of weeds. The task of weeding becomes easier once the soil has been loosened and hence is mostly done by men. The data is also suggestive of surplus male labour. The acreages under the new high-yielding varieties of transplanted paddy, however, are on the increase as government records indicate. This is bound to create new job opportunities for female labour in these regions (Directorate of Agriculture, Madhya Pradesh 1982).

By and large, female participation is very high in weeding operations even if it is predominantly shared with men. The true dichotomy between the north and the south in female work roles may be seen in the highly significant singularly female, input in sowing and weeding operations in the cotton region *vis-à-vis* the wheat region.

Women work at par with men in harvest operations. Cotton picking is almost entirely dependent on female labour. New channels of female work are opening up with the introduction of hybrid cotton wherein the entire cross-pollination operation of the long-stapled American cotton (*Shankar Char*) and the indigenous short-stapled disease-resistant variety is done by women. Here is an example of women's emerging role in new agricultural technology. Unfortunately, diffusion of the new variety is slow due to a low adoption rate among the farmers. Consequently, the government is compelled to restrict production of new seed which pushes the price up, further inhibiting adoption. Binding of harvested wheat in sheaves is overwhelmingly shared by women. Women also share high responsibility in storing of grain.

CASTE AND CLASS IN FARM WORK

Economic status of family is known to be a strong factor on female visibility in public places. Studies document definite shifts in female labour force participation, particularly in the areas of Green Revolution as prosperous farmers have replaced family labour with hired labour (Nath 1965). The opposite process of downward economic mobility and landlessness has also happened with the implementation of the land reform laws. A worsening agrarian economic climate in the small farming community has brought about a greater female input in agricultural production. The data validate increasing female participation with declining holding sizes.

Although caste is constitutionally disavowed as a social phenomenon, data on caste was collected on the premise that the social ethos of caste is very

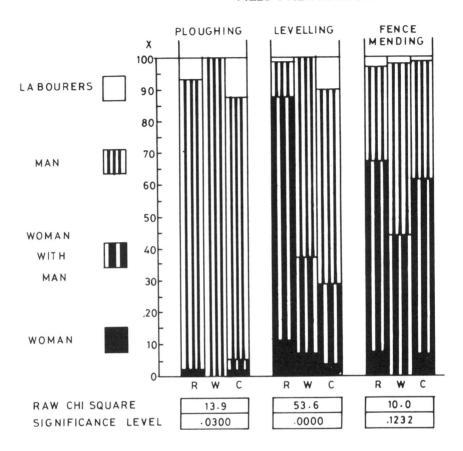

Figure 5.5 Women's work in farming by crop regions, Madhya Pradesh, India

Source: Survey data by author.

SOWING OPERATION

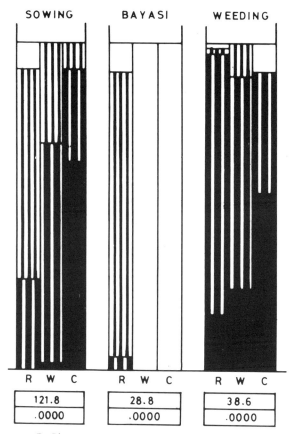

R= Rice   W=Wheat   C= Cotton

151

HARVESTING AND STORAGE

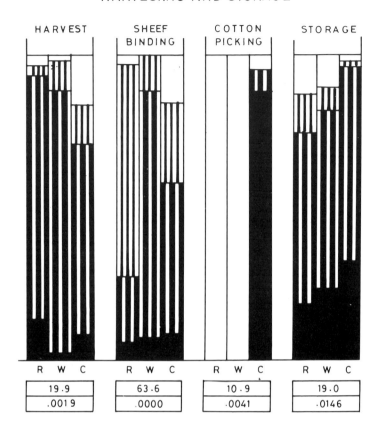

Figure 5.5 (continued)

Table 5.3 Female work participation in farming by caste

| | A | | | | B | | | | C | | | |
| | Stage 1 Field preparation | | | | Stage 2 Sowing | | | | Stage 3 Harvest operations | | | |
Caste groups[a]	M[b]	F	JT	L	M	F	JT	L	M	F	JT	L
1 High	67.5	2.5	22.5	7.5	40.0	12.5	37.5	10.0	30.9	0.0	51.2	17.9
2 Middle	51.7	5.3	37.5	5.5	11.0	30.1	55.3	6.0	25.0	7.5	60.7	6.3
3 Low	36.5	6.45	55.6	7.4	24.3	18.8	53.8	3.2	18.5	7.2	66.2	8.1
4 Tribal	37.5	7.0	53.0	2.5	8.2	37.4	54.1	3.2	17.5	7.5	68.2	6.8

Source: Adopted from D. Bagchi, Female Roles in Agricultural Modernization; An Indian case study, *WID Working Paper Series #10*, 1982: 14.

Notes: a Percentage values are aggregates for each activity.
   b M = Male labour, F = Female labour, JT = Joint Male & Female labour, L = Hired labour

much alive in rural India and must be taken into account in evaluating women's place in a given socioeconomic spectrum. The data confirms the prevailing social ethos of a rather restrictive activity in farming for women of the higher castes. These women are relatively less involved in farm work than the women of the lower-caste groups. Findings disprove a remarkable degree of difference, on the other hand, in levels of female input between the tribal and the non-tribal low-caste group (Table 5.3).

POST HARVEST FARM RELATED WORK

Women continue to work through the post-harvest grain-processing operations. Threshing and husking continue to be manually done, women taking turns with men by conducting the threshing operations during the day while men carry it on through the night. It is women's job to winnow (*udoni*) the grain, often taking half to a full day's work in season. These long seasonal hours drop down to an hour or less for hulling of rice or grinding of flour which can be carried out piecemeal during the entire year as and when time permits. What is noteworthy is that the work carried out within the female space ignores the economic and cultural stratification (Table 5.4). What is also significant is the fluid definition of domestic work for rural women. Men participate in the processing operations so long as these are conducted outside the domestic space. But, as the farm work – cleaning, hulling and storage of grain – moves into the domestic space, men tend to relinquish the responsibility to women. The same is true of livestock and dairy-related work. The animal shed is often located within the domestic space by choice designed to facilitate tending of the herd by the household's womenfolk. Thus, bathing, nursing, feeding of the herd and milking are all exclusively female tasks. Collection of fodder is more of a joint male–female work.

153

Table 5.4 Daily workhours devoted to flour grinding/rice hulling (percent of cases reporting)

	N	< 1 hour	>1 to <4 hours (half day)	>4 hours (full day)
Regions				
Rice	38	71.1	29.9	–
Wheat	59	40.7	59.3	–
Cotton	22	50.0	45.5	4.5
Economic Group				
Small holders	54	52.7	46.3	–
Medium holders	53	49.1	50.9	–
Large holders	12	58.3	33.3	8.3

Source: Survey data by author, 1983–4.

However, the livestock sector – an important ingredient of the intensive farm economy – is almost wholly managed by women, albeit with pride and care. To women, the draft and dairy animals symbolise the family's prosperity and well-being and hence utmost care is accorded to this sector. An interesting anecdote is worth mentioning. In times of acute energy crunch, women prefer to leave their dwellings in darkness rather than the animal shed which must get the best in a system that is heavily dependent on animal power. Ironically, women's perception of themselves is not in the same range despite carrying heavier than normal burdens for their family's sustenance and well-being.

A word must be said on rural electrification and development of appropriate technology. Electrification of the countryside has helped in the creation of mechanised processing. Flour-grinding mills have sprung up in all but the remote villages. Oil-extraction mills are fast replacing the home-extraction process. Commercially produced vegetable oil has replaced the domestic production of clarified butter. Mill-made cloth has replaced home spinning and weaving. All this must be adjudged as progress because it relieves women from the drudgery of manual labour, offering greater leisure time. But, mechanisation and commercialisation of home-based processing in the village has also dealt a blow to women's traditional service roles and income-earning opportunities. The results of my research clearly revealed a widening gender gap, greater social stratification and erosion of female status as a consequence of village electrification and mechanisation. In the same vein, had appropriate time- and energy-saving technology been made available (to women), the opposite could have been hoped for.

SUMMARY AND CONCLUSIONS

Rural women's work is classified in this chapter under 'household' for the domestic, and 'extrahousehold' for the paradomestic. A conceptual diagramatic model of 'Chakra' is used to depict the household work sphere of women. The circular model is to convey a sense of a wheel in motion for the household economy, balancing with the other wheel that signifies the market economy. The suggestion is also of a central role for women in this economy as well as a symbolic confinement with the space that delimits availability of options for women.

In the household sector, provision of fuel has emerged as major work in a situation where acute scarcity of fuel is caused by widespread deforestation. Routinely, women must trek for miles to search for fuel substitutes such as twigs, branches and crop residues. For the majority, the activity is a daily chore involving half to a full day.

Cowdung used to be an inexpensive and easy alternative, plentifully available in the village and easy to collect. But now, the high demand for dung by the affluent population due to the introduction of biogas results in its use as a raw material to produce Methane gas. The substitutes, for which women must trek far and wide all through the year for their daily cooking and heating needs, burn poorly lengthening their work hours and causing great physical discomfort apart from being energy inefficient. Further challenge is posed by increase in family sizes and rising expectations and demands of the family which also enhances fuel needs in that more fuel is required to provide light for children's study, hot water for baths and frequent consumption of tea. The women of the lower income group face a greater dilemma because their time for work is being eaten up in collection of fuel. Redress to their plight is far from sight despite the State's programmes on renewable energy and energy-efficient technology for the countryside. The state's social forestry project, meant as a long-term solution to fuel scarcity, is still in its infancy to produce a substantive dent in the lives of these women, and only a fraction of the population are reaping the benefits of the state plans.

Women's 'extrahousehold' work relates to various stages of farming from field preparation and sowing to harvest and post-harvest operations. One activity that precludes women from participation in farm activities is the use of the plough. The north–south cultural polarity is visible in higher levels of participation by women in the cotton-growing districts of the south *vis-à-vis* the wheat-growing districts of the north. Women of the lower income group are much more active in farm work than their counterparts in the upper income group. Also, in the former group, little variation exists in the levels of female input between the tribal and the non-tribal population.

As the crop-processing activities move within the domestic space, women emerge as solely responsible for the operations of threshing, husking, winnowing, cleaning and storing of produce. Likewise, tending of livestock is primarily women's work, seen as an extension of their nurturant role in the family.

I am hesitant of drawing a generalised conclusion on the impact of new technology on women's work. In some cases, a new technology is labour displacing, producing adverse impact on women's work roles, status, and self-esteem. In others, like in the cases of the hybrid cotton or new rice varieties, new opportunities are created for acquisition of new skills and renewed opportunities of employment. Village electrification, likewise, has rescued women from gruelling manual labour in processing of grain, but has also unsurped their income-earning opportunities. One request that frequently came forth from women during my field work was to help them find something worthwhile and profitable to do during the long agriculturally slack season.

Since this research was conducted, the state government has become more responsive to issues affecting women. A new human resources unit is in place to address the phenomenon of the changing gender roles. It is a worthy beginning, but yet has miles to go. The traditional patriarchy, historically reinforced by feudalistic and colonial modes of production, and a layered bureaucracy are strongly resistant towards any reordering of gender dynamics involving power and privileges. Evidences, however, were clear as I was made aware of a counter force emerging from the oppressed group in the form of a grass-roots movement, this time spearheaded by local women's solidarity groups (*Mahila Mandals*). A more forceful avenue would be the pressure exerted by the professionals, intellectuals and academicians for a redefinition of work in the context of women. Here, a few select recommendations of the Asian Regional Conference of Women and Household held in India in 1985 regarding women's work are pertinent and hence are excerpted below as endorsements:

a) All work involving skills and energy and resulting in a product should be counted as economic. A separation, however, could be made between production and pure maintenance types of work. Domestic activities need to be broken down into i) resource provision, ii) family care, and iii) household maintenance.
b) Production work could be further categorised as i) market oriented, and ii) non-market household production.
c) Market production could be further classified into i) modern paid, and ii) unpaid traditional.
d) Non-market household production could be divided into i) production for cash, ii) for consumption, and iii) for family care. Methods must be devised, however tricky, to distinguish between wage work and self-employment.
e) Finally, in data collection, household approach should be replaced by

gender approach. As much as possible, efforts should be directed towards eliminating bias in data collection of the interrogator as well as the respondent.

This might appear to be a radical overhaul of the existing measurement system, but the time for such a change is ripe.

6

WOMEN'S WORK IN A CENTRAL INDIAN VILLAGE

A photo essay

Doranne Jacobson

INTRODUCTION

In the cool predawn darkness, the soft sounds of turning grinding stones float out from behind earthen walls. Singing quietly as they grind wheat for the day's bread, the women of Nimkhera village in the heart of Madhya Pradesh have begun their day's work.

In this village of some 700 people nestled at the foot of a hill in Raisen district, both men and women work hard at a variety of tasks. Here, members of twenty-one Hindu castes and five Muslim groups farm the village's fertile fields, raising abundant crops of wheat, pulses, rice, fruits and vegetables. Many villagers also work as artisans and construction workers, while some village men are involved in trucking goods throughout this fast-developing region.

The women of Nimkhera village – and the entire Bhopal area, of which it is a part – practice traditional purdah observances. They limit their physical mobility, modestly veil their faces when strangers and elder in-laws appear, and generally behave as decorously as possible. A few practise strict seclusion, emerging from behind household walls only on rare occasions.

Despite limitations on their independent activities, most women of the village work long and arduous hours daily. Although their work is often undervalued by others, and even by themselves, their labour is essential to the well-being of their families and of the larger society.

As in most of India's villages, Nimkhera residents range greatly in socio-economic status. While some wealthy women command the labour of household servants and field workers, others work on their own small farms, and still others toil at menial tasks in the homes and fields of more prosperous villagers. Women's tasks also vary according to age and status within their households. Women who live in joint families often cooperate and divide tasks among several household members.

The accompanying photographs illustrate the scope and variety of women's work in this primarily agricultural settlement.[1]

158

DOMESTIC WORK

Family and child care, household management, food processing and home maintenance

Women often refer to their work as *chula-chakki karna*, i.e. cooking and grinding. These activities are seen as symbolic of the core of women's work, which for all women revolves around food processing and care of the household and children. Some women do only housework, while others perform a variety of other tasks both in and outside the home. Housework and nurturing are ideally seen as the sphere of female activity. Girls are trained in these tasks from an early age.

A child's mother is responsible for attending to its daily physical needs until the child is about 8 or 10 years old. Feeding, washing, supervising, carrying and loving the child are a mother's prime duties.

Providing the household with water and fuel is women's responsibility, as is the preparation of food for consumption. Women carry heavy headloads of water from the well twice a day to their homes, clean the cowsheds and shape the wet dung into patties to dry for fuel, and some trek into the forest to search for wood. They flip grain in winnowing fans, expertly toss out chaff, and painstakingly pick specks of dirt and chaff from platters of grain. They grind wheat in handmills – although some now send their grain out to nearby petrol-powered mills – and spend long hours beside an earthen stove in a hot kitchen, producing delicious meals for their families. Cooking is one of the most important and time-consuming tasks in which women engage.

Milking cows and buffaloes is usually done by men, and women churn the milk into butter and buttermilk. Women are also responsible for scouring dishes and pots, usually with clay and water, and for cleaning the cooking area with an application of yellow mud to cover all greasy spots.

Mending and simple sewing are a woman's task, although stitching of garments is usually left to the village tailors, a husband and wife team.

Women clean their homes daily, sweeping dirt from the earthen floors of the interior and the verandah. Repairing monsoon-damaged house walls with plaster of earth and straw is women's work, as is applying cleansing cowdung paste to the floors and whitewashing the newly-plastered walls. Women also shape new earthen stoves, handmill covers and decorative panels for their homes. Many women paint their walls and doorways with colourful designs.

OUTDOOR WORK

Food production and agricultural and construction labour

Women's participation in outdoor work is conspicuous and essential, with some women working on their own lands and others as paid labourers. However, there are virtually no agricultural or other outdoor tasks which are considered primarily women's work. In nearly every aspect of food production, food gathering, animal herding, and construction of building and earthworks, women are regarded as assistants to men. In fact, arduous outside labour is avoided, if possible, by women of means. Further, there are some outdoor tasks which no women perform.

Women do not drive a plough nor do they drive bullock carts, trucks or other vehicles. They do not take prime responsibility for planning construction or climb on roofs to lay roof tiles.

Women join men in chopping up lumps of earth in fields. At sowing time, they walk beside men driving ploughs, pouring grain into sowing funnels which supply seed to the furrows. They crouch beside men weeding rice and vegetable crops, and work with men to harvest grain with hand sickles, hauling home sheaves in headloads. A few women of prosperous tractor-owning families have tentatively learned to drive a tractor, but only for short distances within the fields. One family owns a large combine-harvester, driven strictly by males.

At the threshing floor, men drive bullocks over the harvested grain stalks, while both men and women winnow grain in the wind. Some women hand-thresh wheat and rice by flailing it or pounding it with sticks.

On construction jobs, labouring women pound earth for roadbeds, haul headloads of mud for mortar and even carry rocks on their heads for building houses and dams. Supervisors are always males.

CRAFTS AND COMMERCE

Artisan and service activities and marketing

Crafts associated with certain artisan castes are often created through the partnership of both men and women. The potter (*Kumhar*) man throws and fires earthen water pots, and his wife and daughters paint them with designs before sale. Potter women also join men in marketing holiday clay sculptures and lamps within the village and at local markets. Similarly, a weaver (*Karera*) man spins cotton threads and fluffs raw cotton for filling quilts, while his wife or sister weaves webbing for beds on a narrow loom. The village tailor (*Darzi*) and his wife sew garments side by side.

Men and women of service castes may also complement one another. The barber (*Khawas*) man and woman each have ritual tasks for patrons' births, weddings and funerals, and the men cut hair, while the women make leaf plates for feasts. Men and women of the washerman (*Dhobi*) caste scrub clothes of patrons of their respective sexes. On the other hand, tanner (*Chamar*) men are called to haul away animal carcasses, but Tanner women do not participate. Sweeper (*Metar*) women work without male help in their onerous task of cleaning latrines and dirty lanes. They also create split-bamboo baskets for sale for household and agricultural use.

Money-making activities not associated with any particular caste are making dried-mud roof tiles, running small shops or tea stalls, and rolling leaf wrappers for country cigarettes on contract.

Marketing is largely the men's realm. Women not observing strict seclusion sometimes visit small village shops, but buying and selling foodstuffs or other items in the regional market towns is men's work. Women appearing in public markets are deemed to be of low status. In contrast, however, women of the very highest status families enjoy shopping for jewelry and fancy clothing in fine urban stores. They are almost always escorted by a family male.

THE FUTURE

Clearly, women's work is vital and integral to the success of village life and will continue to be so far into the future. As mechanisation proceeds, women will probably be all but excluded from controlling the machines, and women are likely to be largely limited to carrying out traditional tasks. In this rural region of Madhya Pradesh, girls' education is growing, but at a slow pace. Unlettered girls of poor families spend their days in traditional menial tasks, while girls of wealthier families may be encouraged to attend school, and, for a few, possibly to find employment in teaching and other fields in the modern sector. This trend may continue until greater social and economic equality is achieved.[2]

NOTES

1 The author conducted anthropological field research in Nimkhera (a pseudonym) in 1965–7, 1973–5, 1979, 1985 and 1991, and is grateful to the residents of Nimkhera for their hospitality and cooperation, and to Miss Sunalini Nayudu for her valuable insights and assistance in the fieldwork. I am indebted to numerous state and district government officers in Bhopal and Raisen for many courtesies extended. The research was supported by the US National Institute of Mental Health, the American Museum of Natural History and the American Institute of Indian Studies.

All photographs are by the author. The photographic prints were processed at the Photo Resource Centre, Springfield, Illinois.

2 For further details on the lives of Nimkhera women, see Jacobson references in bibliography.

Figure 6.1 Inside their courtyard, women of a joint family clean dirt from grain.
Women of a family often cooperate in their work.
©Doranne Jacobson

Figure 6.2 Bringing heavy pots of water from the well twice daily is a woman's job.
©Doranne Jacobson

Figure 6.3 Some women search the nearby forest areas for firewood and carry it home on their heads. As loggers catering to urban demands compete with villagers for forest products, the forests are rapidly becoming depleted, and finding firewood is becoming more difficult.
©Doranne Jacobson

164

Figure 6.4 Shaping cowdung into patties to dry for cooking fuel is a morning routine for most village women.
©Doranne Jacobson

Figure 6.5 Although she is blind, this older woman contributes to the household by grinding wheat in a handmill.
©Doranne Jacobson

Figure 6.6 A mother and her visiting married daughters prepare *simmayyā*, a spaghetti-like Muslim festival food.
©Doranne Jacobson

Figure 6.7 Churning milk from her family's buffalo, a woman makes butter.
©Doranne Jacobson

Figure 6.8 Plastering the floors of her house and veranda with cowdung paste on a regular basis keeps this woman's floors clean and smooth. It is considered proper only for women to plaster with cowdung.
©Doranne Jacobson

Figure 6.9 Two sisters-in-law in a joint family whitewash the walls they have freshly
plastered.
©Doranne Jacobson

Figure 6.10 With her face veiled from the eyes of nearby male workers, a woman helps her husband sow the wheat crop. She pours handfuls of grain down a funnel which deposits the seeds in the furrows ploughed by her husband. ©Doranne Jacobson

Figure 6.11 Bending under the weight of a heavy sheaf of wheat, a field worker takes home her share of the day's harvest.
©Doranne Jacobson

Figure 6.12 A woman of the potter caste decorates the earthen water pots her husband has made. The pots are in constant demand by all villagers. Her craft is traditional to her caste.
©Doranne Jacobson

173

Figure 6.13 Another woman following a traditional caste-associated craft is this sweeper woman weaving a split-bamboo basket. Sealed with clay or dung, the basket is to be used for holding grain.
©Doranne Jacobson

Figure 6.14 A weaver caste woman earns small amounts of money by making roof tiles of mud and straw. The tiles are fired and then sold to villagers for roof repair.
©Doranne Jacobson

Figure 6.15 A weaver woman also practises her traditional craft of weaving cotton cot webbing. She and her brother create most of the webbing used in the village.
©Doranne Jacobson

Figure 6.16 Past an age when purdah considerations are paramount, this high-caste widow runs a small shop in the village. Her earnings are a welcome addition to her income from the small amount of land she and her son inherited.
©Doranne Jacobson

Figure 6.17 In sharp contrast to the women of Nimkhera village, in New Delhi members of the women's wing of the National Cadet Corps march in the Republic Day Parade.
©Doranne Jacobson

Figure 6.18 Also in New Delhi, a woman police officer confers with a colleague while on duty. In contemporary India, women's roles are expanding dramatically.
©Doranne Jacobson

7

India
J16
J21
R23

G10

AGRICULTURAL DEVELOPMENT AND WORK PATTERN OF WOMEN IN A NORTH INDIAN VILLAGE[1]

Huma Ahmed-Ghosh

A cultural interpretation of women's paid and unpaid labour outside the home is presented here by analysing the impact of agricultural development on work patterns of women in Palitpur, a village in the Meerut district of Uttar Pradesh in India.[1] With the advent of the Green Revolution, this village and its surrounding region began experiencing the benefits of agricultural development during the mid-1960s. The introduction of high-yielding seeds, chemical fertilisers and efficient irrigation facilities ushered in a substantial increase in agricultural productivity through large-scale commercial cultivation of crops. Productivity of wheat increased in Palitpur and, due to agricultural innovations, sugar cane emerged as a major cash crop after 1965.

Agricultural development in Palitpur is visible in tractors, tubewells, the use of high-yielding seeds and chemical fertilisers. Development is also apparent in the construction of a tarred, motorable highway running through the village, diversification of wealth into *bhattas* (brick kilns), a medical clinic,[2] a government-run primary school and *pucca* (brick) houses. In sum, Palitpur is recognised as an agriculturally 'developed' village.

Research on women's involvement in wage labour has been inexhaustible. Most research on working women or women with access to an income, links employment to a higher status in terms of female welfare and autonomy. The underlying assumption is that since women have an income, they have control over certain economic resources which they may then manipulate in negotiating their socioeconomic status within the household. This assumption has been summed up by Bennett in her report on women's work in India's agricultural labour force:

> Whether one views the increase in female agricultural laborers as an indicator of growing rural poverty or as a positive sign that more

agricultural work is available for women, its effect in terms of our third criterion (i.e., decision-making) should be examined separately. Evidence suggests that increased paid employment outside the home may actually improve women's bargaining position within the family.

(Bennett 1989: 40).

Such a bargaining position, however, may not necessarily be the case among all classes, especially in households where, despite women's employment, resources may be so limited that equitable allocation and control over them is not an issue. Further, the kind of employment available to rural women is so low in the hierarchy of occupations that women and men engage in it only due to dire economic necessity. Female wage work, especially in the fields, is also seen as weakening of their social status since it is a reflection of the household's poverty. Under such conditions, it is not surprising that women will withdraw from agricultural labour as soon as the household's economic situation permits.

Further, gender hierarchy is so rigid within the household that mere wage labour of women does not guarantee a shift or balance in power relations between genders. Sharma's study of Punjab and Himachal Pradesh appropriately points out that women's wage labour in itself is controlled by household members and not by the women themselves. Sharma explains:

In theory we might certainly expect to find that women who work for wages (and even women who work as family labourers) have a greater say in household matters than women who perform domestic work only, and this is an assumption that has often been made both by anthropologists and others. But the female labourer usually earns wages which are too small and sporadic to lend her any special leverage in household politics, and the work of female family labourers does not give women any particular control over the products of their labour.

(Sharma 1980: 196).

This brings us to another very important consideration regarding women's wage work: the *cultural context* in which particular gender relations are defined and how this affects patterns of women's work. Within the wider social structure, gender relations in India continue to be defined by the dominant patriarchal system. Such a system is the foundation of the pervasive ideology of caste, class and gender hierarchy which has successfully permeated down to the lower castes and classes and is even apparent at the household level – which is the primary nexus in society.[3] Given the existing ideological constraints of a patriarchal system, one finds that gender differentiation is defined within the household itself. This differentiation is responsible for women's subordinate position to the extent that her resources – labour, reproductivity and values are controlled by the males in the households (Sharma 1980).

Women's culturally defined inferior position in the gender hierarchy continues to be one of the most important obstacles in their access to wage labour, education, health services, decision-making powers and, ultimately, any attempt at an egalitarian relationship within the household. Women's roles in society continue to be defined by childbearing and rearing and household duties. A recent World Bank report claims that:

> [It is] part of a pervasive gender ideology which affects the kind of work women seek and the kind they are considered suitable for. It affects inheritance patterns and, thus, the kind of productive assets available to support self-employment. It also affects families' willingness to invest in educating their daughters to prepare them for the job market and, more generally to endow them with the 'bureaucratic know-how' they need to cope with the increasingly complicated administrative structures involved in gaining access to social services and economic opportunity on the 'outside'.
>
> (World Bank 1989: 3).

The dominant cultural ideology prevalent in Palitpur is no different from the ideology perpetuated by the wider social system. Derived from a combination of Brahmanic and upper-class norms, its basic tenets emphasise gender segregation in the private sphere and invisibility in the public sphere. Women should observe purdah, be *pativrata* (serving the husband to the point of worshipping him), obedient daughters-in-law, and devoted mothers and their rightful place is considered to be the home. Such cultural 'constraints' are unfortunately aspired to by lower-class and caste rural women. For lower-caste women, emulation of higher-caste women's lifestyle is an indicator of a high social status. In withdrawing from wage labour and confinement to the house, lower-caste women are also aspiring to a higher economic status since it is a reflection of the household's economic wealth as well as liberation from hard physical labour.

For the purpose of this study, 86 households comprising 106 women were selected out of a total of 229 households in Palitpur. This sample is based on a proportional representation of landed and landless households as well as all caste groups. The latter have been categorised into two groups: higher castes[4] and lower castes[5] based on a rough estimate of their economic position within the village. Higher castes own 69 per cent of the land and lower castes own 5 per cent. The remaining land is either held by non-residents or is community land. Women of the sample are further grouped by three age cohorts: a) younger generation (15–29 years), b) middle generation (30–44 years), and c) older generation (44 and above). All households in the sample are nuclear families. Eight higher-caste families are part of extended families but not joint families since separate kitchens are maintained by them.[6] The socioeconomic consequences of agricultural development on women's work patterns is discussed next. The subsequent section evaluates women's work based on empirical time-allocation studies and the final section discusses women's perceptions of development and its impact on their work patterns.

SOCIOECONOMIC CONSEQUENCES OF AGRICULTURAL DEVELOPMENT

Increased prosperity in Palitpur from the mid-1970s allowed a few higher-caste landowning families to diversify their wealth by investing in *bhattas*. The establishment of nine *bhattas* in Palitpur transformed the nature of the labour force by attracting agricultural wage labourers, both male and female on a large scale. Most of these labourers belonged to the lower castes. In order to understand this displacement of labour from agriculture to *bhattas*, it is important to note that *bhatta* owners paid wages in cash and in accordance with the Minimum Wage Act which were higher than the agricultural wages. This made a dramatic difference to their monetary earnings since agricultural labourers were still paid a combination of cash and kind. After the passage of the Minimum Wages Act, landlords were required to pay wages in cash, but these wages continued to be grossly inadequate. In most instances, agricultural labourers, especially women, continued to be paid in kind and were made to work long hours. The organised and salaried nature of *bhatta* work with fixed working hours made it a more 'respectable' occupation with higher status for its workers as compared to agricultural labour. The permanent (non-seasonal) nature of employment in *bhattas* was also preferred since one of the major causes of poverty and unemployment in rural India is the seasonal nature of the agricultural sector. A labour shortage in the agricultural sector in Palitpur was thus triggered when lower-caste men and women started seeking wage employment in *bhattas*.

Another factor contributing to labour shortage in the fields was the introduction of primary education in Palitpur. As a consequence of economic prosperity and modernisation, higher-caste and lower-caste families started emphasising education for their children. Among higher castes, however, daughters are usually withdrawn from school after completing grades 5 or 8,[7] while sons are encouraged to go on for a college degree. After completing their higher education in the cities, few men choose to return to the village to manage their farms or work on them. Most prefer salaried jobs in urban centres in lieu of the hard physical labour required in farming. A city job also offers glamour and prestige.

In a number of instances, young men who took up employment in cities left their wives and children behind in the village to provide support for their parents and to help out on family farms. Faced with a shortage of male hands in the family and the gradual withdrawal of lower-caste labourers to *bhattas*, higher-caste women are forced to work in the fields. While most older-generation higher-caste women express pride in the fact that their sons are highly educated and are working in the city, middle- and younger-generation women do not readily share this sentiment as they are the ones burdened with extra work. Further, the absence of their husbands also enables the mothers-in-law to exercise an even greater control over the young wives.

Some lower-caste men have also taken advantage of education since they are privileged under a preferential quota system.[8] For the lower-caste community literacy means urban employment, which in turn leads to emancipation from the discriminatory socioeconomic situation rooted in the village. Lower-caste women share the same views as men in their castes and feel that both boys and girls' should have equal opportunities to education.[9]

Lower-caste men prefer city jobs as well as it also enables them to move away from the caste-ridden environment of the village. Further, Palitpur's proximity to Delhi, and the availability of regular transportation has enabled some lower-caste men to either migrate or commute to neighbouring cities for better paying and more prestigious jobs. For instance in Delhi, which is only 40 miles from Palitpur, most of these men have found work as sweepers or as *chowkidars* (security guards). Compared to agricultural labour, these jobs offer better wages and also provide living quarters and a uniform.

Once lower-caste men leave the village, they try to shed their caste affiliations and attempt to move up the caste hierarchy in the anonymity of the sprawling metropolis. A common strategy for attaining upward mobility is the adoption of a different family name – usually one belonging to that of the middle-range higher caste group. At the same time, village-bound lower-caste women try to raise their social status within the village community by withdrawing from wage labour, observing purdah and maintaining a social distance from working women in their caste group. A few lower-caste men also withdraw from wage labour in the fields to work on their own land which they acquired from the government under the *Bhoodan* movement.[10]

Rapid agricultural development and economic prosperity in Palitpur have thus altered the work patterns of women in the village. Eighty per cent of higher-caste women from the sample were forced out of the confines of their homes and into the fields to engage in arduous agricultural work. Meanwhile, 34.61 per cent lower-caste women from the sample abandoned agricultural wage labour for work in *bhattas* while 34.61 (coincidental) per cent withdrew totally from the labour force. The outcome of this change in work patterns may be better understood through a detailed description of women's time-use. The purpose of such an analysis is not simply to describe women's daily routines but also to highlight women's preferences in their use of time and the obstacles to better employment of time. More importantly, patterns of women cannot be understood unless a detailed recording

184

of all their activities, economic, productive and household, is attempted. Since household activities are difficult to record by conventional methods of census, surveys and interviews, time-allocation becomes an important tool in the analysis of such data (Minge-Klevana 1980). Work patterns of women in Palitpur have been analysed by collecting data on the time allocated for various activities. The specific activities selected for recording are: outdoor work (wage and non-wage), livestock care and household work.

TIME-ALLOCATION ANALYSIS OF WOMEN'S WORK

Management of time for women is an important determinant in their quest for a higher social status and must be given adequate attention in any attempt to improve the socioeconomic position of women. Since women are primarily responsible for the maintenance of the household, they are left with little time or flexibility to pursue skill development or training for more lucrative and 'respectable' jobs. Another constraining factor, though not in its entirety, on women's free time is the lack of innovation or improvement in household technology which would reduce time spent in housework. Cultural insensitivity and deliberate perpetuation of a gender hierarchy by men are other contributing factors to the lack in easing women's time-consuming and monotonous work both inside and outside the home.

OUTDOOR LABOUR

In Palitpur, higher-caste women belonging to the younger and middle generations spend 5 to 5.5 hours daily outside the home working on their family farms (Table 7.1). Women now help in sowing and weeding, tasks which they previously did not perform. These new tasks are in addition to their regular duties of threshing, cleaning, storage of grain and complete care of livestock. Higher-caste women belonging to the older generation are not significantly involved in agricultural work, averaging only 2.81 hours a day. However, their help is indispensable during the harvesting season.

Nearly 34.61 per cent of lower-caste women in the sample who previously worked in the fields of higher-caste farmers have moved to *bhattas*. Most of these women belong to the middle and younger generations. These women spend more time at work outside the home than the older generation, mainly because they are physically stronger and are burdened with greater financial responsibilities to support their families. On an average, they spend over six hours outside the house working for wages.

LIVESTOCK CARE

With agricultural development and the increase in wealth, purchase of livestock has risen significantly in Palitpur. Livestock rearing has always

Table 7.1 Total number of hours spent on outdoor labour per day

Caste groups	Non-wage agricultural	Wage agricultural	Wage bhatta
Higher[a]			
15–29 (n = 15)	5.14		
30–44 (n = 13)	5.33		
45+ (n = 12)	2.81		
Lower[b]			
5–29 (n = 7)		7.00 (n = 2)	6.49 (n = 5)
30–44 (n = 5)		7.10 (n = 1)	6.38 (n = 4)
45+ (n = 5)		6.00	

Source: Based on survey by the author.

Notes: a From the sample three women each from the 15–29 and 30–44 age groups and nine from the 45 and above age group did not engage in agricultural work outside the house.

b Among the lower castes four women from the 15–29, five women from the 30–44, and three women from the 45 and above age groups did not engage in outside agricultural labour.

been an integral part of higher castes' subsistence. Prior to agricultural development (pre-1965) in Palitpur, milk and dairy products were not sold beyond the confines of the village and were primarily consumed by the family. Now, the large number of cattle owned is not only a symbol of wealth, but also the modus operandi for extra income. With the increase in livestock has come the sale of milk and dairy products to neighbouring villages. Though it has succeeded as a profitable venture, the additional work to run it is shouldered exclusively by women from higher-caste families. Lower-caste families own some cattle but not in sufficient numbers to earn a living from them. On an average lower-caste families own one or two buffaloes.

A time-allocation analysis of women in Palitpur reveals that livestock rearing requires substantial effort, on an average accounting for 4–5 hours per day (Table 7.2). Higher-caste and lower-caste women collect fodder while engaged in other agricultural activities. They take some time off in the afternoon to collect fodder, then return to work in the fields and *bhattas*. For lower-caste women, there is added pressure from their employers to get back to work as soon as possible. Lower-caste women spend roughly 1–1.76 hours a day on fodder collection. The landlords allow lower-caste women to

Table 7.2 Time spent by women in livestock activities per day

Caste groups	Collect fodder	Cut and feed fodder	Collect and make dung cakes	Milk cows	Other*	Total hours
Higher						
15–29 (n = 18)	1.36	1.35	1.00	0.25	0.28	4.24
30–44 (n = 16)	1.83	1.38	1.20	0.28	0.26	4.95
45+ (n = 21)	–	1.66	1.05	1.78	0.11	4.60
Lower						
15–29 (n = 9)	1.66	1.06	0.43	0.25	0.08	3.48
30–44 (n = 9)	1.76	1.05	0.50	0.33	0.23	3.87
45+ (n = 8)	1.00	0.89	0.75	0.50	0.50	3.64

Source: Based on survey by the author.
Note: *Others include bathing cattle and cleaning the cattleshed.

collect fodder from their fields but keep half of it as compensation for their generosity!

Since higher-caste households possess a larger number of livestock, they spend more time in the care and milking of cattle than lower castes. While higher-caste women do spend more time than women from other caste groups, the mean total number of hours per woman devoted to livestock care does not reflect a wide range of variation because lower-caste women spend a few hours collecting and cutting fodder for the higher-caste households.

HOUSEHOLD WORK

The retreat to the confines of the home has been gradual but deliberate in patriarchal societies. Household chores (cooking and cleaning) and child-rearing are the focus of women's activities (Raju 1991). In a sense, the home and especially the kitchen, is one place where women in Palitpur control and wield power. However, this study found that in Palitpur development has reversed the expected benefits among the caste groups. Higher-caste women who go out to work on family farms spend relatively less time on household work. Despite their wealth, higher caste women spend nearly the same amount of time as the lower castes on household work.

Higher-caste women spend approximately 2.5 hours a day on cooking. Even though they cook twice a day, they tried to keep the time spent in the kitchen to the minimum. The longest cooking time was reserved for dinner, which was their major meal. This meal took about 1.75 hours inclusive of the time spent in serving food to men in the family.

For lower-caste women most of the day is spent in wage labour hence not much time is left for household work. Little time and effort is spent in preparing meals. In part, this is due to the fact that lower-caste families do not earn enough to indulge in elaborate meals. Lower-caste women spend 1.4–2.7 hours per day on cooking. Dinner is the principal meal for lower-caste households also, but unlike the higher castes, it is usually a meager one. Since the custom of serving food to men first is not widespread among lower castes, time spent on cooking and eating is minimised.

Table 7.3 shows that most higher-caste and lower-caste families depended heavily on well water for their cooking and cleaning. Higher-caste women of the younger and middle generations spent 45–8 minutes per day collecting water, while their counterparts among the lower castes spent 60–5 minutes per day, mainly because of the scarcity of wells in the lower-caste section of the village. Although the act of fetching water itself may not involve much time, social exchanges add up to make it a full-fledged activity. Time spent on cleaning activities is nearly the same for all castes and age groups, not because their houses are the same size, but because some lower-caste women spend time cleaning the outside of higher caste houses and remove their garbage. For higher-caste women, cleaning the house involves sweeping the courtyard and the one or two rooms the house has.

CHILDCARE

Quantifying time spent on childcare was not a straightforward task. It was difficult to distinguish between time spent on childcare and time women spent on relaxation in the afternoon because of an overlap of the two activities. Childcare is not considered an exclusive chore as it is never perceived as an activity which requires special attention or extra time. It is therefore performed simultaneously with numerous other activities.

According to Table 7.3, higher-caste women of the younger and middle generations spend the least amount of time on childcare (28–40 minutes). This is not due to any neglect of the children but because older members of the extended family help in looking after them. In most cases it is the older women who spend 1.5 hours per day in childcare. Also, all higher-caste children are away at school for most of the day, allowing their mothers to go to work.

Since most lower-caste members live in nuclear families where an extended kinship network is absent, they are unable to leave their babies with

Table 7.3 Time spent on household chores by women per day (hrs)*

Caste groups	Cooking servicing	Fetching water	Cleaning activities	Childcare	Total
Higher					
15–29 (n = 18)	2.73	0.80	1.72	0.66	5.91
30–44 (n = 16)	2.46	0.75	1.53	0.46	5.20
45+ (n = 21)	1.72	0.00	0.50	1.50	3.72
Lower					
15–29 (n = 9)	1.97	1.08	1.26	1.30	5.61
30–44 (n = 9)	2.07	1.00	1.32	0.61	5.00
45+	1.41	0.33	1.25	0.41	3.40

Source: Based on survey by the author.

Note: *Table only lists the hours spent on household work by women who engage in outdoor labour. Women who do not engage in labour on an average spend 7.50–8.00 hours a day on household work, with cooking and childcare taking up the bulk of women's time.

elders. Lower-caste women usually carry the baby to work. When the youngest child reaches the age of 2, it is left in the care of older siblings. Hence, the time women do spend with their children is while feeding and bathing them, or when the children accompany them on visits to neighbours.

The above description of women's work patterns reflects the heavy workload women have to shoulder. The time-allocation analysis also highlights certain activities which are time consuming and contribute to the family economy but are not considered productive by men in the family or census enumerators, specifically livestock care and the production of dairy products. Further, time spent in fetching water or in lighting a stove can be cut down if appropriate technology is available.

WOMEN'S PERSPECTIVES

This study also incorporates data on women's views of the impact of agricultural development on their work patterns. Such an analysis is crucial to any socioeconomic study because it takes into account women's own perceptions of their status, lives and needs in the village. Perceptions not only reveal what women think about a particular situation, but are also a reflection of their socialisation and their internalisation of gender relations as projected by society at large.

Analysis of the responses of higher-caste women reflects a discontent with the effects of economic development on their work patterns. Higher-caste women do not deny certain benefits of agricultural development, especially since their families are wealthier than before. They claim that economic wealth is apparent in their daily life – they wear better clothes and eat marginally better. Yet higher-caste women feel that it has not led to 'appropriate' modernisation where women can access better educational opportunities and health facilities, or aspire to greater autonomy, independence and greater decision-making within the household. What agricultural development has brought about is increased work, without recompense in cash, kind, prestige, or appreciation from their family and community. Subsequently, increased work has deprived them of what little leisure time they previously enjoyed.

Mostly, higher-caste women resent working in the fields. They point out that farm work involves tiring physical labour and is not an occupation with high status. In a community where family status is tied to confinement of women to the home, purdah and non-involvement in any visible form of labour except livestock care, farm work is a departure from tradition and definitely not a welcome one.

Discussions with higher-caste women reveal that older-generation women are acrimonious towards lower-caste women and blame them for the present occupational structure in the village. The older women complain of the government pampering lower castes and not taking interest in the plight of higher castes. Higher-caste women of younger and middle generations are less critical, but remain unhappy with the current situation. Going to the fields daily means rising early, rushing with the housework and then spending most of the day in hard toil. Pushpa, a 19-year-old Chauhan woman claims that development has brought wealth to the village and farmers but not to women. Pushpa continues,

> We now have to get up early, eat fast and rush to the fields. When we come back in the afternoon, the cattle have to be tended to. We are so tired. We neither get wages or recognition from the family for all the work we do nor do we have the freedom to do what we would like to.

Since all the labour performed by higher-caste women is on their own family farms, they do not draw remuneration. In keeping with traditional economic logic, therefore, it is not even considered 'work'. The prevailing male attitude in Palitpur subscribes to this view as well, because they too deny that women in their household do any work outside the confines of their homes. Higher-caste men refuse to acknowledge the agricultural work done by their women because they consider it a poor reflection of their own socioeconomic status as providers.

190

According to higher-caste women, being in control of livestock and dairy products does provide them with some power and sense of worth in the family. Higher-caste women in Palitpur at no point complain about time spent in livestock rearing. However, although women are responsible for the welfare of livestock and decisions on the quantity of milk and dairy products to be sold, they are not directly involved in the sale of it. This is due to cultural restrictions on women's mobility and social interaction with males outside the family, especially with members of other castes. Women display a strong sense of conformity to tradition and do not feel that this denies their 'rightful' economic role. Just control of livestock appears to provide them with enough pride and appreciation from the family that a level of self-esteem is maintained. But such self-esteem is not displayed when they respond to questions about agricultural labour and its impact on their position in the family and community.

Higher-caste women do complain about not spending as much time in the kitchen as they did before. They perceive the kitchen as women's domain and believe that through the cooking and serving of food they wield some power in the family. They see their role in the family being diminished when this activity is shortened and becomes less elaborate. Sushama, a 32-year-old Brahman woman pointed out that:

> My mother spends a lot of time in the kitchen cooking elaborate things like sweets and snacks. I too would like to do it, but here I do not even get enough time to cook a decent dinner. Neither do I get enough time to eat my meals at a relaxed pace.

On the contrary, lower-caste women are positive in their responses to the impact of development on their lifestyles. They claim that the establishment of *bhattas*, a *pucca* road and education are direct outcomes of development. As stated earlier, development in Palitpur is seen by the lower castes as an opportunity to move out of the repressive socioeconomic whirlpool in which they are caught. While for lower-caste men development has translated into migration to cities for better economic and social opportunities, their female counterparts have reacted to this situation by expressing an eagerness to adopt the customs of the Brahmanic tradition with as much fervour as their economic position will permit.

Lower-caste women of the younger and middle generations are also grateful for development in Palitpur because, in many instances it has led to their withdrawal from wage labour altogether. These women express pride in their confinement to the house and some of them even compare their 'fortunate' position to that of the 'unfortunate' one of higher-caste women who are forced to work on family farms. Some of the lower-caste women are aware of the labour shortage that their shift away from agricultural labour has created and believe that it reflects the progress their community has made.

CONCLUSION

The concurrence of social, political and economic changes arising from agricultural development has contributed to a change in the work patterns of women in Palitpur. The economic changes that involve increased wealth and livestock, in conjunction with access to education has led to a reversal in the work patterns of women. It was observed that higher-caste women are working longer hours than they did before, and lower-caste women are shifting to *bhattas* or withdrawing from wage labour altogether.

Household work, childcare and animal husbandry are activities esteemed by all women in Palitpur. Agricultural labour, which is lowest in the hierarchy of occupations, is the least sought after. *Bhatta* work is hard but ensures steady wages and is therefore preferred over agricultural labour.

What emerges from discussions with the respondents on issues relating to women's employment pertains not only to the social and structural constraints created by a male-dominated society, but also to their own acquiescence to a subordinate role. Women belonging to households at subsistence level do not distinguish between 'productive' labour and household work and claim that 'all the work they do is for the family'. The reasons for this have been aptly stated by Bhattacharya who asserts that the asset base of households at subsistence level is so small that most of the production is for household consumption. Bhattacharya maintains that:

> Because total production work input of household members will be much less in such [subsistence] households as compared to that in households having a larger asset base, the contribution of women members of households as participants in generating this non-marketed product can only be small in absolute size, however crucial it might be to the survival of the household. The separate identity of this marginal work, which in major cases is performed only intermittently together with women's other work of housekeeping, is more often likely to be lost in reckoning. In effect the work becomes 'invisible'.
>
> (Bhattacharya 1985: 200)

As women's wage labour, caste and class positions are inter-related, employment outside the home has different connotations for women of different castes and classes. For women engaged in manual and 'low status' jobs, the nature of work, its position in the occupational hierarchy coupled with low wages does little to improve their position within the family or community. Women from extremely poor lower-caste families suggested that a factory or a small industry needed to be set up in the village so that women could seek employment. On the other hand, women from landless but not impoverished lower castes claimed that they would work at home. As Jain points out:

[In north India] certain cultural modes are a severe barrier to the demand pull for labour. Even poverty does not push them into outside work to the extent that it does in other parts of India. . . . In other words, the push of poverty overcomes the constraints of culture if mediated by the right type of 'secluded' work.

(Jain 1985: 222, 238)

Women from higher caste households were also keen to work but a reluctance to work outside the house was also apparent. Surprisingly, the wealthier lower-caste women expressed no desire to work. This illustrates the inter-relationship between cultural and attitudinal constraints and women's status.

Women have limited access to agrarian technology and rural development programmes. This exclusion is not commensurate with their role in agriculture as producers and processors. While women's contribution in agricultural labour is proportionate to that of the men, their efforts go unrewarded. This is reflected in the exclusion of women from most decision-making processes and direct access to agricultural technology. They are rarely given the opportunity to learn about improved farm methods, technological skills, cooperatives and livestock activities. Credit facilities are less easily available to women, sometimes because loans are given against land titles which are issued in men's names. Therefore, if rural development is intended to benefit women, it is imperative to ensure greater access of women to agricultural technology, rural institutions and extension services. Greater participation of women in development schemes which assist them in real economic terms such as control over land, access to loans and training in modern methods of cultivation could change their perception of agricultural labour. It should be noted, however, that such assistance has to go hand in hand with an attempt at cultural adjustments leading to a shift in 'perception' of women's wage work by both men and women.

In Palitpur, agricultural development has quite clearly resulted in economic wealth for most households, but its effects have to be seen at an intra-household level. For higher-caste women in the village, development is more apparent in the high workload some of them have assumed than in a higher literacy level, better nutrition and increased medical facilities. Women's labour power is not only tied to the households' socioeconomic status but is also controlled by men in the family. Changes in the occupational structure have led to the emergence of higher-caste women from their homes, which could have only been possible with the sanction and insistence of men in their families. However, an economically induced shift in patterns of women's work along this line is still not the norm. This is apparent in the resentment towards such work by higher-caste women and the denial by higher-caste men that their women engage in outside labour.

Not only has agricultural development drawn higher-caste women out of the confines of their homes into agricultural labour, but it has also led lower-caste men and women into other higher paying and respectable jobs like work in *bhattas* or in neighbouring cities. What is apparent from my research in Palitpur, is that 'work' which is physically hard, culturally demeaning, devoid of economic remuneration and control over economic resources is not preferred by women (or indeed, by men). Women withdraw from such labour as soon as the family's economic situation improves. Women who do work at wage or non-wage labour do not achieve independence, power or even participate in decision-making processes regarding agricultural and major family decisions. Thus, if 'work' by women does not lead to autonomy, gratitude from family males, recognition and acceptance from the wider community and attention from policy makers, most rural women would rather not engage in it. Given the sociocultural aspirations of rural women in Palitpur, it is imperative that rural development incorporate women's views, perspectives and social status. It is only in this manner that a more balanced model of development can be achieved and an attempt at social change be made.

NOTES

1 Palitpur is a pseudonym for the village.
2 Though the men in the village boast of the existence of a clinic, I rarely saw a doctor there. The doctor was based in Baghpat (closest town to the village) and was operating another clinic there.
3 See, e.g., Saraswati Raju, chapter 2 of this volume.
4 The higher castes, Brahmans and Chauhans are grouped together because they belong to the same economic class. While the few Brahman families in Palitpur do perform their ritual duties in the village, their main occupation is as landlords.
5 The lower-caste comprise of households belonging to the Shudra and Untouchable castes and include *lohar* (blacksmiths), *darzi* (tailors), *chamar* (sweepers), *nai* (barbers), and *bhangi* (untouchables: toilet cleaners and carrion carriers). While most members of the lower castes engage in agricultural and *bhatta* work, each caste group also performs its caste-ascribed occupations.
6 The extended families in the sample are treated as nuclear in much of the analysis because while the members of these families reside in the same house, each nuclear family occupies a partitioned section of the house.
7 An offshoot of agricultural development and economic prosperity has been the adoption of the dowry by the residents of Palitpur. Educating daughters implies marrying them to educated men, which in turn involves a dowry, the amount of which increases with the level of education of the man. Hence, higher-caste families refrain from educating their daughters beyond a certain grade to avoid paying exorbitant dowries.
8 The Indian Constitution guarantees equal opportunities in the achievement of education and jobs to all citizens irrespective of caste, religion and gender. Some lower castes are further protected in the Constitution by a quota system which is based on a 20 per cent reservation for scheduled caste members in all government schools, institutes and employment places.

9 Unlike higher-caste women, lower-caste women were keen to educate their daughters because a) some of them were willing to pay the dowry for an educated groom, and b) some of them wanted their daughters to have 'respectable' jobs (like teaching, nursing and secretarial) in the city.

10 The *Bhoodan* movement was part of the 1970 Land Reform Act. Under this Act, landless agricultural workers were allotted land given away by landlords and big farmers as a voluntary gesture to cultivate for themselves. Most of this land was barren and on the outskirts of the villages.

8

DIVISION OF LABOUR AND WOMEN'S WORK IN A MOUNTAIN SOCIETY
Hunza Valley in Pakistan[1]

Sabine Felmy

INTRODUCTION

Mountain societies are known for their remoteness and backwardness. A rigid social structure with little scope for social mobility and substantial women's participation in decision-making are often attributed to these communities in South Asia. On the other hand, the environmental setting of the mountains requires a close cooperation between all household members to make use of their scarce and vertically stratified resource potentials. This results in close-knit communities where everybody knows one another. These communities also lack a social veil like purdah and the space of women's activities is wider as compared to most women in other ecological surroundings. They are further characterised by dependence on subsistence farming catering to home consumption. Production of cash crops and marketing are limited because of lack of opportunities. Of late, however, modern changes and increased accessibility through a network of roads are responsible for an exchange pattern which influences not only the mixed mountain agriculture – an interdependent system of crop raising and livestock keeping – but also the gender division of labour, the lifecycle of women and their role at different levels of society.

In this case study of the Hunza Valley in the Karakoram (administered as part of the northern areas of Pakistan), women's participation and their growing responsibility in the domestic sector is analysed with emphasis on their changing position in Hunza society. An overall socioeconomic transformation results in the reorganisation of group positions, especially in areas where out-migration of male workers is increasing, as is the case in the valley. The Hunza Valley was selected for the study because of dramatic changes that have occurred with the construction of the Karakoram Highway. The construction of the road was followed by infrastructural development in rural areas with an aim to improve the living conditions and speed up the process of nation building. The study examines the uniqueness of the Hunza society, and the impact of development on gender roles and on women's work.

THE ENVIRONMENTAL SETTING

Completed in 1978, the Karakoram Highway was built by Pakistan with the help of the Chinese over a period of twenty years and connects the Hunza Valley with China. It follows the Indus and Hunza rivers, crosses the border on top of Khunjerab Pass at a height of 4,700 metres and leads into the autonomous region of Xinjiang.

Apart from its strategic position, Hunza has always been an isolated high mountain region in the Karakoram until the end of the nineteenth century. Isolation also meant independence. From the twelfth century AD a dynasty of *mir* or *tham* ruled the Hunza Valley. The capital of their kingdom was Baltit situated at an altitude of 2,438 metres. The former palace which was the residence of the *mir* until 1948 can still be seen towering above the old village.

Mirdom and independence survived attacks from adjacent principalities and enemies from across the borders. Peaceful negotiations with China and Kashmir helped to keep up a status quo of regional self-administration. However, peace was seriously threatened in 1891 during the famous 'Hunza Campaign' when British India sent an army in anticipation of a Russian invasion and installed a buffer zone. *Mir* continued to rule but was supervised until 1948 by British 'political agents' residing in Gilgit Town, the capital of the Northern Frontier.

Profound changes have occurred since 1974 when the then president Zulfikar Ali Bhutto terminated independence and *mirdom* in the Gilgit Agency. Consequently Hunza was administratively integrated into Pakistan and was connected by a metalled all-weather road, the Karakoram Highway. Hunza now has an undecided political status as the inhabitants known as 'Hunzukuc' do not pay taxes, get subsidised food stuff from 'down country' (a regional term for the low-lying areas of Pakistan, mainly Punjab, Northwest Frontier Province and Sind) and remain unfranchised.

The term 'Hunzukuc' covers four different ethnic groups, i.e. Burusho, who occupy central Hunza and as the speakers of Burushaski form the major ethnic group; Wakhi settlers in the northern part of the Hunza Valley known as Ghujal; Shina speakers who dominate the southern section of Shinaki; and the Dom – traditionally blacksmiths and musicians – who inhabit only one village (Mominabad) near the Hunza capital Karimabad (formerly Baltit). In spite of many different ethno-linguistic groups the people of this region have a fairly common material culture based on irrigated agriculture and livestock.

Every study on Hunza has to deal with two aspects: a) the local conditions, and b) the position of Hunza as a part of Pakistan. So far the

197

Hunzukuc lived mainly on agriculture which is a time-consuming occupation in an area with steep, barren mountain ridges and little rainfall (130 mm per annum on an average). In this mountain desert surrounded by nearly 8,000-metres-high peaks and vast glaciers, melt waters from tributary streams are the only source for field irrigation. They alone secure village life in small oases of 30,000 people.

The daily struggle for survival requires commitment and group effort. Apart from permanent water shortage, the Hunzukuc are always threatened by the wrath of nature. They live in close contact with nature and have learnt to make the most of its resources. They are energetic, hard working and witty people with a social system steeped in the feeling of family and community.

In Hunza women can be seen everywhere: in the streets shopping and carrying baskets (*giran*), in the fields and gardens where they spend several hours every day and at meetings and festivals. They wear baggy trousers, long shirts, nice embroidered caps (*cuke pharcin*) and no purdah is observed. Instead, they cover their head – but seldom their faces – with a *dupatta* (scarf). A characteristic feature which distinguishes Hunza from many other areas of Pakistan is that most of the Hunzukuc have been Ismailis – followers of His Highness, the Aga Khan – since the beginning of the nineteenth century. The present Aga Khan IV introduced many development projects providing health and educational facilities to this community. Hunza, as a predominant Ismaili sub-division, benefited from these innovations which contain special arrangements for women. The Girls' Academy in Karimabad was inaugurated in October 1987 by His Highness himself. This boarding school for 350 female pupils marks a new era of women's participation in social life.

COMPOSITION OF THE HUNZA HOUSEHOLD

An analysis of women's role has to focus on the household. In Hunza, a household has an average of 6.2 members. Often three generations live under one roof, i.e. grandparents, parents, sons and unmarried daughters. These traditional extended families form a self-contained unit combining residence, production and consumption. They live in a one-room house with a fireplace in the middle, and two platforms each for sitting during the day and sleeping during the night. The smokehole in the flat roof above the fireplace is the only opening other than the entrance door. In the back of the house is a storage room, and corners of the dwelling contain cupboards, shelves or boxes (Figure 8.1).

There are two major differences between the Burusho-house (central Hunza) and the Wakhi-house (upper Hunza). The two platforms for sitting and sleeping are opposite each other in both housetypes, one is for men (*uyum man/kla raz*), the other one only for women (*jot man/nisine raz*). In

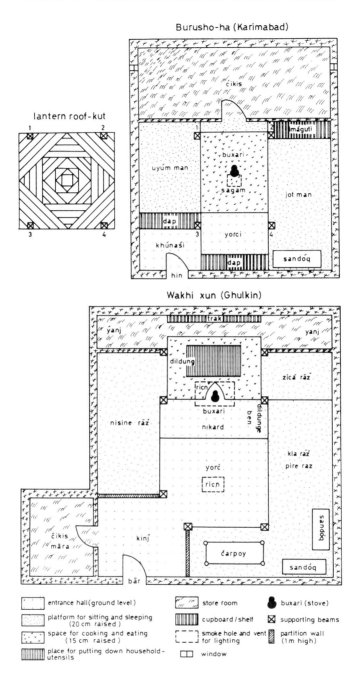

Figure 8.1 Types of houses in Hunza (ground plan)

Source: Kreutzmann (1988).

the Burusho-*ha* (house) women sit to the right of the entrance and the Wakhi women sit on the left: the platforms of high esteem are reserved in both house types. Visitors always have the privilege to sit next to the fireplace on the men's platform. For cooking and serving women stay behind the fire or *buxari* (stove). In the Wakhi-*xun* (house) the women's domain in the back of the room is open where there are shelves and cupboards, whereas the area behind the cooking place in the Burusho-*ha* is walled up and closed by a door and contains the treasury of the house under the authority of the eldest housewife. Thus, the store-room functions also as a refrigerator for keeping meat as central Hunza is warmer, due to its location down in the valley. On top of the roof is a place for drying fruits, vegetables and grain and a veranda where the whole family lives from May to October.

The one-room house and closely knit family system, however, is breaking up and is in some cases already reduced to joint families with brothers and their families sharing one roof but running separate households. Changes have occurred due to migration, off-farm occupation and education. Accordingly, the houses' plans have also changed as extra rooms are being attached to the traditional *ha* for guests, young couples and for each son and his family.

The extended family system maintains a social hierarchy. Age commands authority and respect which helps to maintain a system of seniority as a natural controlling force in the composite household (Müller-Stellrecht 1979: 157).

The eldest working man has the highest authority and responsibility. He decides on buying or developing land, on buying and selling or exchanging cattle, grain and other property. He looks after the agricultural equipment and deals with surplus products (unground grain, butter, cloth) and money. The 'master of the house' is the only one who knows about the hiding place of *maltas* (rancid butter) which is often buried for years as a special treat for festivals or family events.

The oldest lady of the house (*ruli gus*) is responsible for the household implements, the food supplies and annual rations. She alone has the keys for the storage bins and the *cikis* (store-room). When the housewife-in-chief is away, no one has access to any supplies. The capability of the eldest housewife is measured by her ability to avoid starvation which in former times was a regular threat during spring. False or insufficient rationing could lead to a family disaster and eventually even to divorce.

The *ruli gus* is treated with respect and her wisdom and competence are held in high esteem. She delegates work to the younger female members of the house. In general, the workload is shared within one household and quite a few decisions are discussed among all the members. Lorimer comments on family life in Hunza:

the favourable attitude towards women, who enjoy a satisfactory status of perfect human equality within their own sphere. They are treated neither with contempt as inferiors nor with exaggerated consideration as valuable subhuman pets. They seem to be regarded as a perfectly good human type with special functions which they will normally fulfill satisfactorily. They are accordingly trusted and left in perfect freedom to go about their business.

(cited in Müller-Stellrecht 1979: 158).

Contrary to the above statement of human equality, a distinct difference can be noticed in the upbringing of boys and girls right from their birth. Sons are not only heirs and guarantors of patrilineal descent but stay in the extended family home with their parents all life long. Daughters move to their husbands' homes after marriage, often at the age of 12 and are more or less lost to their parents as a working member of the household. They have to accept their husbands' families and their new role as the daughter-in-law with special duties under the direction of the matriarch. They become accepted members of their new families only after giving birth to a child. Overall, girls learn much earlier than boys to take over responsibility. When a son is born, there is a gunfire from the roof of the house to announce the happy occasion followed by celebrations. At the birth of a girl there are no congratulations but the mother's parents will give her a cooking vessel or pot as a bridal trousseau. Boys stay in the cradle for two years, girls only for one. Boys get weaned after three years, girls after two. Both sexes are also treated differently with regard to fasting: girls begin as early as at 9 years old while boys not until the age of 14.

Children are involved in the daily tasks from very early on. Girls look after their brothers and sisters when they are old enough to leave the cradle; they sew or embroider and collect twigs, thorns or hay. Boys mind the cattle, milk goats and help with the irrigation of fields and gardens. However, in Hunza nowadays nearly 80 per cent (boys and girls) attend a number of state and Aga-Khan-sponsored schools.

The school-going children help in the household chores in the afternoons by taking care of grazing cattle which is essentially done by boys while girls fetch water, look after livestock, dry fruits and cook. Harvesting and threshing are jobs performed by both boys and girls. Girls learn to bake bread at the age of 10 and regularly cook when they are about 12 years old. They leave home for marriage when they are conversant with their basic duties.

Contrary to the Islamic laws, land and other heritage is normally divided among the sons to avoid further splitting up of small households. After the parents' death girls inherit no land but their brothers buy them some jewelry and clothes, grain and household utensils on the day of their wedding. Since the turn of the century there have been some alterations in the inheritance rule and girls are allowed to inherit if there are no male off-spring of the deceased.

DIVISION OF LABOUR IN A HUNZA HOUSEHOLD

Writing as far back as 1939, Lorimer observed that:

> There is no hard and fast rule about what is women's work and what is men's except that, for obvious reasons, the extra heavy loads are handled by the men; but a man will gladly take a turn at driving the threshing team or carrying the baby, and a woman will readily do a spell with the winnowing fork or shovel. No job is taboo for anyone able to tackle it, and the result is the pleasantest possible family co-operation.
>
> (Lorimer 1939: 102–3)

What was registered in the 1930s can still be observed in certain cases although there have always existed different duties for men and women (Table 8.1). It is of particular interest to analyse shifts in this organisational structure which can show the degree of adjustment to altered economic conditions in Hunza. 'To sit in the mountain grazing ground' is listed under men's work by Lorimer. This is correct for central Hunza but in upper Hunza (Ghujal) women go to the high pastures together with their children. In other parts, however, they are not even allowed to set foot in grazing grounds situated at high altitudes as they are considered to be impure not only during menstruation time or after giving birth, but all the year round (Kreutzmann 1989).

Everywhere men used to cultivate summer pastures. Nowadays, most of them are giving up this for jobs as porters for mountaineering expeditions and other non-agrarian activities. Once it was a privilege to go to high pastures with livestock and to produce milk products which are important winter supplies for the whole community. Several men took care of the village livestock as a whole and got paid in grain and milk products. Now only a few old men volunteer to go to the high pastures and are also a common sight in agriculture since most of the able-bodied men migrate or take up income-generating jobs.

There is little ideology behind the division of labour except practical reasons. Normally men do the physically heavy work, jobs which demand long absence, or take a longer time. Women's mobility is limited, especially when they have children. As such, they have to stay near the house, cook food, tend children. In certain areas purdah is observed which implies that women travel only in the company of their husband or a male relative.

There is evidence about one major difference between men and women concerning job-sharing. That is, women work twice the time per day as compared to men. It was noticed by Lorimer that:

Table 8.1 Traditional division of labour in Hunza

Men's work	Women's work	Joint tasks
• to irrigate	• to scatter manure or sand on the fields	• to reap, thresh the crop
• to clean out irrigation channels/ponds	• to pick up stray grains	• to turn cut crops
• to cultivate the soil	• to sweep the threshing	• to beat buckwheat
• to carry dung	• to use winnowing tray	• to cut the ears of millet
• to tie up the cut crops	• to fetch baskets of grass	• to dig up potatoes
• to tie cows together for threshing	• to tie up grass in bundles	• to pull up turnips and onions
• to put the grain into sacks	• to grow vegetables	• to sit at the mill
• to pound odd ears of corn	• to plant potatoes	• to bring in refuse and cowdung
• to act as miller	• to rub cowdung on the trees	
• to grid the year's supply of grain	• to gather apricots and mulberries from the ground	• to sweep up dead leaves
• to put the flour into bins	• to milk the cows	• to pluck out goat's hair and tease it
• to cut grass	• to prepare flour for grinding	• to lay out warp and make bobbins
• to store dry leaves	• to take small quantities of grain to the mill	• to carry shoulder-baskets
• to milk goats	• to control the baking of bread in the house	
• to shear sheep/goats	• to dry greens/apricots	
• to sit in the mountain grazing ground	• to split apricots, break fruitstones	
• to make buttermilk	• to extract the kernels	
• to tie up butter and put it in store	• to wash kernels and pound them	
• to climb trees	• to extract oil from them	
• to plant saplings	• to make syrup	
• to keep guard over vineyards	• to bring in salt-earth from the desert and extract salt	
• to make wine/spirit	• to sweep the house/lane	
• to spin goats' hair	• to put down bedding	
• to twist warp yarn on the big spindle	• to wash clothes/cooking utensils	
• to weave cloth from wool and goats' hair	• to beat wool and prepare it for spinning	
• to slaughter animals and skin them	• spinning	
• to knead skins, make skin sacks	• to twist warp yarn on the wheel	
• to make skin coats	• to wind yarn	
• to build house	• to knot warp for weaving woollen cloth	
• to fix thorns on the top of walls	• to sew, patch, embroider	
• to fell and bring down firewood	• to make ties for boots/socks and coats	
• to split firewood	• to make loops for whips and wedding tassels	
• to cook meat/gruel for a feast	• to carry hand-baskets	
• to make spoons, rolling pins, bread-turners, cooking pots from stones kneading trays	• to sit at the sluice	

Source: Müller-Stellrecht 1979: 160–2; and own observations.

> In Hunza from ancient times a man who is able to, does every kind of
> men's and women's work, and a woman who is able to, does both
> women's and men's work. So if a woman sits down, they say she is a
> man. . . . There is the saying: 'Is this female a man or is she a woman?'
>
> (cited in Müller-Stellrecht 1979: 160)

In recent times, outmigration of male workers from Hunza has increased
dramatically. The pattern of sharing work as outlined above in a Hunza
household is put under considerable stress due to lack of manpower.

To make up for this deficit various measures are undertaken by every
household. Wherever possible hired labour or mechanisation is used.
Formerly only men used to carry manure to the fields (Table 8.1). Now both
sexes do this work, though labourers are hired to carry the dung baskets.
Where machines can reach the fields, threshing operations, which earlier
involved the labour of the whole family, are performed by Chinese threshing
machines. Nevertheless, a large portion of the extra work is shouldered by
women who not only make up for lost male hands but also for school-going
children who are no longer available for work. The reduced number of
helping hands in each family has also resulted in increased support from
neighbours and members of the patrilineages.

A recent research report by the Aga Khan Rural Support Programme
(AKRSP 1987) revealed that for the busiest time of the year in Gerelt, a
village located in central Hunza, women work fifteen hours per day in
August which is the month of wheat and apricot harvest including housework as well as cutting of grass and alfalfa. They work fourteen hours on
average in September when they thresh, clean and grind wheat, cut grass and
alfalfa, and weed maize and millet fields.

The longest working days, i.e. eight hours for men are in August and again
in November with the ploughing of wheat fields and the pruning and
planting of fruit trees. Gerelt is a double cropping area where wheat is
followed by maize. There is little livestock in the village. The average size of
landholdings is 0.4 hectare. Men look for off-farm income and employ
coolies from Baltistan as farm hands. Women do most of the agricultural
work including carrying manure from cattlesheds to fields if enough coolies
are not available. Jobs women do not engage in are ploughing, planting and
pruning fruit trees, and construction work. Notwithstanding increase in
household income, the workload of women in the Hunza Valley shows a
significant increase. This is in contradiction with a survey in rural northwest
Pakistan where rise in income is associated with a reduction in women's
weekly working hours (Hippel and Fischer 1987: 12).

The time schedule for the month of August for one central Hunza
household in Gerelt is provided in Table 8.2 as an example of the labour
constraint.

The household consists of ten members. Two women and one man work

Table 8.2 Labour constraint in one household of Gerelt (central Hunza)

Time	Female head	Daughter-in-law	Male head
04:00	pray	pray	pray
05:00		make tea	
05:30	breakfast*	breakfast	breakfast
06:00	harvest wheat	harvest wheat	harvest wheat
07:00	harvest wheat	harvest wheat	harvest wheat
08:00	harvest wheat	harvest wheat	harvest wheat
09:00	harvest wheat	harvest wheat	harvest wheat
10:00	harvest wheat	harvest wheat	harvest wheat
11:00	harvest wheat	cook food	harvest wheat
12:00	harvest wheat	harvest wheat	harvest wheat
13:00	lunch	lunch and wash up	lunch
14:00	harvest apricot	harvest apricot	harvest apricot
15:00	split apricot to dry on roof	split apricot to dry on roof	cut firewood
16:00	harvest wheat	harvest wheat	harvest wheat
17:00	harvest wheat	harvest wheat	harvest wheat
18:00	collect fodder	collect fodder	harvest wheat
19:00	feed livestock	pick vegetables	irrigate potatoes
20:00	dinner	dinner	dinner
21:00	store dried apricot	wash up dishes and fetch water	irrigate
22:00	store dried apricot	wash up dishes and fetch water	irrigate
23:00	sleep		sleep

Source: Timing adapted from AKRSP 1987: 22.

Note: *Preparing breakfast also involves collection of twigs for fuel, fetching of water and preparation of *phiti* in the evening of the previous day.

in agriculture. The small holding of 9 *kanal* (0.45 hectare) is split into parcels of wheat followed by maize and a perennial crop of lucerne. Five cows, two goats as well as . . . twenty-four fruit trees and twelve poplar trees for construction and firewood belong to the household. The three workers each spend 6 hours in the morning and another 6–7 hours in the evening on agricultural work. Changes are occurring in the Hunza Valley due to the construction of the Karakoram Highway which has increased outmigration. But even earlier, although remotely located, Hunza has always demonstrated flexibility rather than a rigid gender approach to the management of the household economy. This rapid adaptation to a changing economy of Hunza illustrates the opposite of the social complex a few miles down the valley in Nager where the division of labour is more rigid and men see women's role as essentially that of childcare and food preparation.

In Hunza, age and efficiency are the main components which determine the division of women's work. The *ruli gus* (eldest woman) in the household supervises all women's work and does jobs which need experience and special skills such as sowing of vegetables, care of the livestock and cooking for guests, etc. She is responsible for food and fodder supplies and has to keep an eye on the monthly rations. Also, the eldest woman may replace a man's work of going to the summer pastures in Ghujal. The *ruli gus* finds spare time for handicrafts often in winter months when most agricultural duties are not required. She does the spinning of the sheep wool while men treat goat's hair and weave rugs and woollen cloth or *pattu* for coats, vests and caps (see Table 8.1).

The eldest woman also delegates to the younger women the preparation and serving of food and stall feeding. The daughters and the daughters-in-law are mainly responsible for cooking, dish-washing, fetching water, washing clothes and general cleaning. They also do needle work especially knitting and embroidery.

All women take care of poultry farming, vegetables, weeding, harvest and drying of fruits. A woman with a baby might carry out more jobs close to the house. When the child is older, she does more field work and looks after the livestock. Thus, there is a clear division by age in women's work and yet flexibility exists if circumstances demand.

The social system of extended and joint families also contribute to the household's ability to cope with the socioeconomic changes. The women are not left alone as in the nuclear families when males take up jobs away from their homes. The system allows optimum use of various resources while demanding a much greater individual effort for the sustainability of the household.

RURAL DEVELOPMENT AND WOMEN'S PARTICIPATION

With the completion of the Karakoram Highway in the Hunza Valley the northern areas had become a part of Pakistan's market economy with a steady flow of goods into the mountain areas. The building of the road was followed by infrastructural improvements such as the building of linkroads, dispensaries and schools by the Public Works Department. Besides, the Food and the Agricultural Organisation (FAO) and the United Nations Development Program (UNDP) sponsored Integrated Rural Development Programmes (1981–5). The AKRSP has also been introducing new methods of agriculture. These organisations focused on land development by extending the irrigated village lands, increasing production of cash crops by introducing improved seed varieties, fertiliser and mechanisation. These strategies are well known throughout Asia as the Green Revolution and the focus is mainly on male farmers. In the first phase, the development organisations decided about village-level projects in consultation with men only. In the second phase separate women's organisations (WOs) were introduced which had a good response only in specific villages where women were already engaged in income-generating enterprises such as quilt-making, carpet-knotting and poultry production. Women themselves favoured technical devices like nutcracking machines, mechanical butter churners, threshing machines, which they perceived as time- and energy-saving devices that would reduce women's workload. In the third phase, women's organisations were integrated with the village organisations and in the overall development activities. An increased sensitivity to their problems is being observed though not a reduction in their tasks.

CONCLUSIONS AND PROSPECTS

Recent changes in the Hunza economy has its impact on the social position of women. As women had to extend their sphere of responsibility beyond the household, they have entered new fields of responsibility. The more women are involved in agricultural decision-making and income-generating enterprises, the higher has risen their social status in Hunza. This is contrary to findings in rural northwest Pakistan where increased income levels have reduced working hours for women placing them in a higher social status. However, it has also meant less integration for women into the socio-economic life of the village (Hippel and Fischer 1987: 12–13).

Because of the women of Hunza, the cultivated lands there present a picture of intensive care contrary to other areas where parcels of fallow land abound because men have left the villages. The dynamic approach to the gender division of labour and the immense effort on the part of the Hunza women combine to infuse the interest of the community in preserving the region's resource potential. Women have not only become farm managers, but also, by virtue of efforts of the Ismaili community, educational opportunities have opened up a spectrum of non-agrarian jobs for urban and rural women. The establishment of the Aga Khan University in Karachi and special scholarships for female students from the northern areas have opened up ways for the local students to come back to the mountains for medical service in their own communities.

At first female teachers were brought into Hunza from 'down country'. Now, Hunzukuc themselves work as female teachers mainly in primary schools. Step by step women's participation in every sector of life is increasing. All of these are part of a strategy to improve the living conditions for women in the future. It is only a beginning and is quantitatively less significant as compared to job opportunities for males. However, relative to other high mountain areas of northern Pakistan this is a big qualitative step.[2]

NOTES

1 The fieldwork for this study was conducted during 1984 and 1985, preceded by shorter visits in 1981 and 1983 and follow up visits in 1986 and 1989. During one year, the agricultural calendar was closely observed with regard to activities.

Gratitude is expressed to friends in Hunza who supported my work, to the staff of the Women's Section of AKRSP, to Prof. Dr Hermann Berger who allowed the use of his unpublished Hunza–Burushaski dictionary and to Hermann Kreutzmann of Berlin who supplied the illustrations for this chapter.

2 The example of Zohra Bibi from an irrigation colony near Gilgit Town where many Hunzukuc families settled shows how income-generating enterprises for women have a spin-off effect for the next generation. Zohra Bibi is the first woman plant-protection specialist who received training by AKRSP (so far only 275 men have been trained in this field). To quote from the report,

> she climbs fruit trees and waters the wheat fields and sprays the crops with borrowed equipment. She earns Rs 250 [about $12] a week. She says that the money from her earnings will go towards the schooling of her eldest daughter.
> (Aga Khan Foundation 1987: #5)

Part III

GENDER IDEOLOGY, POWER AND POWERLESSNESS: EMPIRICAL OBSERVATIONS

9

TOWARDS INCREASED AUTONOMY?

Peasant women's work in the North-Central Province of Sri Lanka[1]

Joke Schrijvers

In 1977–8, I carried out policy-oriented anthropological research on the situation of peasant Sinhalese women in the North-Central Province in Sri Lanka. The general aim was 'to describe and analyse the situation of women under changing social and economic conditions, and to give recommendations for a development policy aimed at the needs of women and their participation in society' (Postel and Schrijvers 1980: i).

In this chapter I will present my image of the conditions of work of the women among whom I lived in Kurunduvila, a *purana* (old, ancient) village in the North-Central Province of Sri Lanka. I lived there for a year, together with my family: my partner (who carried out research as well) and two children of 5 and 7 years old.

Instead of continuing 'participant observation' and interviewing, I decided during the second half of my field research to give priority to action-research involving a group of women belonging to poor families. By organising a 'collective women's farm' I assisted them in improving the conditions of their lives. The last section of this chapter deals with the changes we aimed at. I use the concept of women's autonomy to structure the analysis.

GENDER DIVISION OF LABOUR

Early in November 1977, after a two-month preparatory period in Colombo, we settled down in Kurunduvila. Academic colleagues in Colombo and officials from different levels and backgrounds in the area had given me the impression that Sinhalese village women were even more restricted in their movements than Dutch middle-class housewives. According to these urbanised views, peasant women in the North-Central Province were still backward housewives busy with cooking and childcare. I did not unquestioningly accept such images although I had little knowledge about women's participation in agriculture. The literature I had studied had

211

given no gender-specific information regarding agricultural tasks. If women were mentioned at all it was mainly in sections on kinship and marriage, as indispensable links between males (Leach 1961; Yalman 1967). I primarily associated the rural women still unknown to me with responsibilities in and around the home.

During my first few weeks in the village, I visited all houses, getting to know the residents and drawing up a simple household census. The village consisted of 63 households with a total of 366 inhabitants; except for one family they were all Sinhalese Buddhists belonging to the *Govigama* caste. My first priority was to get acquainted with women, and to find out what their activities were. I had difficulties, however, in meeting them at home. It was early November, the peak of the main agricultural season (*Maha*). Except for the very old, the very young and some unmarried girls, the village was empty during the day. Girls, young mothers, middle-aged and older women worked in their *hen* (*chenas*, swiddens), or they were engaged in casual labour, miles away from their homes. I chatted with the women in their kitchens during early mornings when they prepared food to take to the fields. I also walked with them to the *chenas* in the surrounding forests. It was absolutely normal and necessary for women of all ages to work in the fields. They took their children with them or left them at home in the care of an elder daughter, old mother or other relative (in one case a baby was left daily with her aged grandfather, as the mother and grandmother both worked as casual agricultural labourers).

The literature available on the region had obscured or distorted the facts about the work and the economic importance of women (Brohier 1975; Ellman *et al.* 1976; Leach 1961; Peries 1967; Samarasinghe 1977; Yalman 1967; Wickremeratne 1977). My own research, however, gradually convinced me that for centuries women's agricultural labour had been essential for the survival of the population. In Boserup's language (1970), one could even speak of a 'female crop': the sowing and reaping of *kurakkan*, finger millet, proved to be exclusively women's work. Men and women rationalised this by stating that reaping this crop was 'too troublesome for men; it was backbreaking and men had no patience for it'. In addition to the work on the *chenas* which was the responsibility of women and men, male cultivators were predominantly responsible for paddy cultivation with the aid of rainwater, collected in artificial lakes (*wew*, tanks). Paddy production in Sri Lanka's Dry Zone is more precarious than the production of millet, maize, gingelly and vegetables on the *chenas* in the forests. In about three of every five seasons the paddy crop is a failure due to drought (Brow 1978: 98), whereas the cultivation of millet requires only a few short showers.

Only two generations ago, all villagers had lived on subsistence agriculture. I estimated that under those circumstances the gender division of labour gave men and women an equal share of work in cultivations. While

212

men cultivated paddy (aided by women and children during harvest time), women carried out the bulk of *chena* cultivation. Rice which is the 'male crop' was the most favoured food. Throughout the centuries, however, the population had survived on millet – the predominant crop from the swiddens. Millet is more nutritious than rice, more drought resistant and less perishable. It can be considered as a 'famine crop' for the women kept it as a foodstock and 'savings-account', to be sold in small quantities whenever they were in urgent need of cash.

Contrary to what people told me about their recent past, the present economy was a mixture of subsistence and market agriculture. There was now an increasing dominance of the market sector with the accompanying need for money. Even the *chenas* were being primarily used for cultivation of cash crops such as chilies, cowpea, soya, green gram, mustard and gingelly. These products were traded by men, the women having done most of the work of cleaning, planting, weeding and harvesting. In the subsistence economy, both women and men bartered some of their products such as gingelly and mustard seeds with traders from the north who came to the villages. In exchange they obtained betel, palm sugar, salt and dried fish.

One important effect of the penetration of the market economy was that women had much less control on products of their labour, and on the income derived from it. The dominant gender ideology prescribed restrictions to women's mobility. They were not expected to drive a bullock cart, a lorry or a bicycle. As such, the changes taking place in agriculture had, almost automatically, transformed the gender power relations in a fundamental way, i.e. women's access to and control over the basic means of production had diminished rendering them much more dependent on the income of their menfolk.

This process was accentuated by the agricultural extension officers who selected their 'contact-farmers' from predominantly the males of the villages along the main road. Women in general and families living off the main road in the old centre of the village were mostly cut off from this direct source of information on 'modern' agricultural methods. To comprehend fully the changes brought about by this 'development trend', it is necessary to understand the historical processes at work described briefly in the following section.

HISTORICAL BACKGROUND

The ancient and medieval centres of the Sinhalese were located in the region that today is known as the North-Central Province. It is generally held that the history of the Sinhalese people in the region began with a wave of Aryan immigrants from northern India around 500 BC under the leadership of king Vijaya. The immigrants established settlements based on a combination of slash-and-burn agriculture on swiddens in the forests and on paddy cultivation. The climate was harsh with seasonal but unreliable rains and frequent spells of drought lasting for years at a stretch. The ancient Sinhalese developed a sophisticated irrigation system which enabled them to grow paddy crops despite the capriciousness of the rainfall. Irrigation was made possible first by tapping rivers and afterwards through the construction of small individual village tanks. Later, these were transformed into an elaborate network of large and small reservoirs with connecting canals (Gunawardena 1973: 5–13; Silva 1977: 32).

At the beginning of the first century AD, after the consolidation of the kingdom, large-scale irrigation works were constructed using water from the Mahaweli Ganga and other rivers whose sources lay in the southwestern Wet Zone. With increasing sophistication of irrigation technology the region became densely populated (Hettige 1984: 30) opening up extensive tracts of land for intensive cultivation.

During the thirteenth century AD, weakened by internal power conflicts and external invasions, the irrigation civilisation collapsed. In the north, a Tamil kingdom was consolidated. The Sinhalese centre of power shifted southward near Colombo, the beginnings of the Kandyan kingdom emerging in the centre of the island. From then on, until the end of the nineteenth century, the north-central region reverted to its original state containing only small and scattered hamlets with sporadic village tanks.

Very little is known of the conditions that persisted in the north-central region for six centuries beginning from the thirteenth. The area was isolated from the centres of power: the Tamil kingdom of Jaffna in the north, and the Sinhalese kingdoms of Kotte and Kandy in the south-west and the centre of the island respectively. The conditions of isolation persisted through the Portuguese and the Dutch colonial rule.

The Kandyan kingdom maintained its independence during the successive colonial occupations. The provinces of Nuwarakalaviya and Tamankaduwa, currently known as the north-central Province, formed part of the dominions of the king of Kandy. The impact of the Kandyan administration was insignificant due to the remoteness of these provinces (Hettige 1984: 30–1). The isolation continued even after the British established control over the Kandyan kingdom in 1815 (Silva 1977: 44–63).

The colonial policy was directed primarily at the development of commercial plantation agriculture in the hill-country areas of the Wet Zone. In the Dry Zone, agricultural policy was oriented almost exclusively towards increased paddy production in the form of cash-crop cultivation. The British considered *chena* cultivation as primitive and unfit and tried to abolish it. But the efforts were successfully resisted by the peasants as was the grain tax introduced in 1818, which was subsequently abolished in 1892 because of the difficulties of collection (Hettige 1984: 50).

The peasants of the North-Central Province were not affected directly by all these measures until the turn of this century when the British surveyed the area and classified land under different categories. This encouraged private enterprises to purchase what had been declared as the Crown Lands. *Chena* land was included in the category of unoccupied waste land (Hettige 1984: 51).

In the course of the present century, peasant settlements and reconstruction/restoration of ancient irrigation works became integral to the economy of the colonial government. By 1920 the opening up of new land was perceived as the most important means to raise food production and to relieve the pressure on land in the Wet Zone. For the first time during the colonial rule the focus shifted to the peasants of the Dry Zone. After 1930, a new settlement policy was directed towards resettling the landless poor from the overpopulated Wet Zone through colonisation schemes in the Dry Zone. This policy continued after independence in 1948 by the successive Sri Lankan governments. But, the total reintegration of the North-Central Province with the rest of the island took place only after 1948 when programmes of large-scale peasant colonisation and irrigation reconstruction became major components of the government's policy (Hettige 1984: 29).

Only since 1950 has the full impact of the changing modes of production been felt by the peasant population of the North-Central Province. At the time of my research, the oldest inhabitants still remembered the forests intersected sparsely by cart tracks, an existence based on subsistence production and seasonal barter, and the 'arrival' of the outside traders, teachers, administrators, roads, buses, shops and cars.

In trying to trace the effects of the commercialisation of the economy, I was struck by the difference in perceptions of changes among different economic groups. The well to do romanticised over the old abundance of buffaloes, milk, fish in the tank and wild life in the forests but also welcomed the change. The less affluent who had to struggle for their survival pointed out the general deterioration of the conditions of 'the poor' (*duppath*) and the emergence of a sharp distinction between the poor and 'the rich' (*pohosat*) cutting across family relationships. Indebtedness had increased leaving poor families with little option other than selling pieces of paddy land to the affluent landowners who were also the money-lenders. The deep emotional effects of this emerging class division on the poor was noticeable. Their loss of access to basic resources such as land, water and forests was felt not as mere economic and material loss, but perhaps even more predominantly, as a social loss. It was their human dignity that had been directly affected.

Equally pronounced was the vigour with which women and especially mothers would analyse the material deterioration of their existence – of increasing shortages of food, cash, clothing, housing, etc. The women did not directly associate their worsening conditions with the transformation in gender power relations that I, as an outsider, perceived from a number of factors. Women, more than men, were cut off from or refused access to resources that were of significance to the new modes of production: private

215

capital, land, labour, new means of transport (lorries and tractors), information on new agricultural technology, credit and development of entrepreneurial capabilities.

POVERTY, GENDER AND CLASS-DIFFERENTIATION

The less affluent women with whom I discussed these recent changes saw their growing poverty in conjunction with their womanhood as an extra burden. The gender ideology prevalent in the region made women see themselves as the weaker sex. Symbolically, the superiority of the male was expressed in different ways. Rice was regarded as the 'male' crop and given a higher value than millet, the 'female' crop. Similar was the differential value placed on male and female work. Although both men and women considered motherhood to be the greatest fulfilment in a woman's life, notions regarding impurity (*killa*) of the female body during menstruation and after birth made women feel inferior to males. Childbirth itself was considered an extremely painful and humiliating experience. The idea of their impurity made women feel vulnerable to evil spirits; it circumscribed their behaviour. Hence even though women wished to become mothers, the physical process and the responsibilities of motherhood, especially in situations of poverty, were perceived as burdensome.

During our conversations, many married women pointed at the differences between their lives and that of unmarried girls. They also pointed out the greater responsibility of women in comparison to men towards their families. 'Men are like dogs', an elderly woman explained to me, 'they go out whenever and wherever they like, and come home only to eat' (to eat, *kanava*, also had the connotation of having sexual intercourse).

Gradually it became clear that the women's own perception of lack of power and control was connected with the sense of their own bodies. Most women over 40 had at least four children but would have preferred a smaller family. It struck me that most women, young or old, literate or illiterate, had some basic knowledge of modern contraceptives though very few used the 'pill'. The men had negative attitudes towards contraceptives. The women blamed their husbands for not permitting them to use the pill or go for sterilisation. For poor women, the daily anxiety of feeding the family was enhanced by their persistent fear of an unwanted pregnancy.

I discovered that the relative degree of a family's poverty was directly related to their access to work in the 'modern' white-collar sector. The population pressure on agricultural resources was creating smaller and more fragmented land units. Families who had at least one member employed as teacher or clerk were structurally more secure than those who had to subsist exclusively on agriculture or on wage labour. The 'modern' sector thus was a big factor in creating economic and social disparities.

Increasingly, poverty forced women and men into working as casual labourers for the local élite. For such families it was difficult to combine *chena* cultivation with wage labour. I noticed that often women would take the embarrassing first step to hire themselves out to affluent farmers in order to augment family income. The prevailing gender division of labour which held only women responsible for feeding their households allowed men room to avoid the ultimate responsibility of the household and also of facing the consequences of increasing poverty.

Apart from the gender relations in general, the relations among women themselves had also changed. The growing distinctions between 'the poor' and 'the rich' now placed women in competition with one another in emerging situations of structural inequality. The earlier system of mutual help and exchange labour characterising village life had transformed into relationships of patronage and matronage as well as in emerging class relations (Schrijvers 1985). In subsistence agriculture, the exchange of labour between different households was the basic organising principle. Women and men would form work parties, *kayiya*, providing mutual help during the peak of the agricultural season. This system could only last under fairly egalitarian conditions where the work shared was about equal for all parties. Increasing inequality of landownership and greater demand for hired labour had broken down the *kayiya* system. This had an adverse effect on the neighbourly relations among the women and their abilities of giving expression to a political counterpoint (Wertheim 1954, 1964). When women worked together in a *kayiya*, during the breaks they would play-act scenes (*sellam karanava*, playing as they called it) reversing the prevailing gender and power relations around, thus venting their criticism of the existing social order. Under the changed economic circumstances of competition for wage labour, there was much less room for this kind of group activity – an institutionalised lighthearted form of rebellion against the dominant value system (Wertheim 1964: 23). Rural commercialisation had breached the conditions which were conducive to female collaboration. It should not be construed, however, that an idyllic pattern of female solidarity ever existed prior to this transformation.

A STRUGGLE TOWARDS AUTONOMY: THE COLLECTIVE WOMEN'S FARM

The more I felt at home in Kurunduvila, the more I became uneasy and unhappy about my research. Carrying out 'policy-oriented' research (aimed at giving recommendations for a better development policy for the government of The Netherlands) seemed to me a detour from stimulating direct solutions. After eight months of fieldwork I experienced 'fieldwork depression'. I felt tired, useless and sick. I was bogged down with questions such as: what was the use of collecting all this information and how could my research become a meaningful contribution for the women I had begun to care deeply for? I had begun to realise how much the conditions of life had worsened in one generation. I was also being approached by women from the extremely poor households with a request to help them with a better and more respectable way to earn income rather than as casual wage labourers. Some of these women suggested the establishment of a collective farm for women. 'Wouldn't it be fair that women should also be given a chance to improve their existence?' I was often asked.

By the end of June 1978, I had formulated a plan that was the outcome of lengthy discussions with these women, agricultural development officers and some successful entrepreneurs in the region. The plan was presented to The Netherlands' embassy in Colombo and later funds from The Netherlands' bilateral aid programme were obtained. Thus, my fieldwork developed into what may be termed as 'action-research'.

At this point, I would like to stress that I learned more during this latter phase about power relations in general and gender relations in particular than in the preceding months. I gained more insights into:

a) the strong solidarity developing between the village élite and the local bureaucracy;
b) the means of power used by the local élite against the poor, particularly women and their reactions; and
c) the barriers to resistance particularly by poor women and hence the necessity of bringing them together through conscientisation and self-organisation.

The first phase of my research was, however, essential for me as an outsider, to understand why these women wanted to change their fate, why they were unable to do so on their own, and why they needed a catalyst to help them find a way.

The conceptualisation of the happenings in the initial years of the women's farm was based on the changes occurring at crucial points that affected the prevalent gender power relations. I later labelled these points as the main elements of women's autonomy, a concept which refers to increased control of women over their own lives and bodies, ensuring in the process a sense of dignity and self-determination (Jayawardena 1982: vi; Schrijvers 1985: 236). My experience in working with the women in Kurunduvila made me realise that the concept of women's autonomy is inseparable in its ideological and material sense. For women to achieve more autonomy, increased self-confidence and a sense of dignity are equally important or, perhaps, a prerequisite to achieve control over material resources. In my view, the main interlinked elements of women's autonomy are:

a) women's control of their own sexuality and fertility; forms of shared mothering between women or between women and men;

b) a division of labour which allows women and men equal access to, and control over, the means of production;

c) forms of cooperation and organisation which enable women to control their own affairs; and

d) positive gender conceptions which legitimise women's sense of dignity, self-respect and their right to self-determination (Schrijvers 1985: 236).

These elements on the one hand indicate a direction for change. They refer to a sociopolitical ideal – a society in which relatively egalitarian relations regarding gender, class and ethnicity can exist. On the other hand, they can serve as analytical tools for evaluating the degree of women's autonomy in a specific situation, culture or period. It is important to distinguish between autonomy as a collective principle and as an individual asset. For instance, it can refer to the self-determination of women's groups *vis-à-vis* the established (political) organisations, but also to individual women who want to achieve a greater say in their relationships within the family.

Coming back to the discussion at hand, in the following analysis of the changes brought about by the organisation of the collective women's farm in Kurunduvila, I use autonomy mostly in its collective sense. The farm, however, also created more individual autonomy for many women members.

Approximately forty women rented a piece of Crown land from the government of Sri Lanka to attempt collective vegetable and fruit farming and marketing. The objective was to increase their earnings, strengthen their gender position and regain their dignity.

With the seed capital from the donor agency, the women could earn a wage for each day's work on their own farm. This made possible participation of even the poorest of women who were in no position to wait for income until after the sale of their produce. The collective organisation made it possible for these women to improve their situation in more than one way. Economically, they acquired control over the product of their own labour and after a year and a half none of the participating members needed work from the local élite and a few were able to buy back a piece of their own land. By breaking through their dependency relationships, they had achieved some degree of social independence. Although these women still needed outside support, they had, as a group, full managerial control over their own project through a committee selected on a yearly basis from the working members which controlled the finances and laid down the policy outlines. The experience also influenced their activities outside the project as they acquired new skills and more general knowledge of the farm business. Politically, as poor women they were able to enlarge their power base to some extent. They became less competitive with each other and much less dependent on the local élite. They also became more conscious of their collective interests as a group and as women. During the

219

twelve years since the project started they became better organised and many members achieved a greater self-confidence. They became visible as a group and audible to the authorities. The main obstacles to progress, however, turned out to be the harsh physical climate on the one hand and the increasingly violent political situation in the country on the other. I come back to this in the concluding section of this chapter.

What is the net outcome of the changes brought about by the collective farm when analysed on the basis of the four elements of women's autonomy identified earlier? Regarding the first element – the degree of *control over their own bodies* – two changes are noticeable. A number of women started to use contraceptives and some opted for sterilisation. The women had some knowledge of the possibilities of contraception but the principal impediment to its adoption was opposition from husbands. Many feared physical violence from husbands. It seems that becoming financially less dependent on men was a factor in women's independent decisions about their fertility. Second, it has a bearing upon their physical mobility which increased considerably.

This element needs further research. It would be interesting to examine whether there is a connection between women's self-determination with regard to their reproductive powers and their spatial mobility. Incidentally, all the younger women have learned to ride bicycles for their shopping or visiting doctors on their own. They have also become independent and more at ease travelling by public transport, for instance, to local offices of the agricultural department.

Regarding the second element, i.e., the *control over the means of production*, the women's farm changed the established gender division of labour. The women began to control their own business thereby reversing the norm of male authority. Moreover, they acquired the 'male' skills that corresponded with prestige and authority. They learned to drive a tractor and to use simple agricultural machinery. Having done this, they have debunked myths about women's lack of ability in certain 'male' fields. Men who belonged to the local élite reacted at first in public with derogatory comments and jokes. When this did not help, a few lamented: 'there is no room now for men in the village, the women have taken over'. The majority of the women paid little attention, and did not give up their new responsibilities. They themselves attributed this to the strength of their 'holding together'. Changes in the division of labour did not mean less tasks in the domestic sphere such as cooking and childcare. With few exceptions, the women continued their household duties as before. A collective creche at the farm site has not become a reality because of more urgent organisational problems. Increase in the productivity of poor women on the one hand and simultaneous alleviation of their domestic burden on the other, therefore, have remained problematic.

As to the third element in the autonomy concept – *forms of cooperation and organisation of women to control their own affairs* – the women's farm enabled its members to increase their skills and improve living conditions. Time and again, women stressed the value of their collective undertaking and their 'being together'. However, there was a flip side to this collective behaviour because some of the poorest women found it difficult to refrain from the extremely competitive behaviour that had been their strategy of survival over the past decades. Therefore, the formal and informal support received from outside groups and individuals remained crucial in guiding and motivating the women in spite of the many difficulties and conflicts.

The situation improved when, in 1986, a small local non-governmental organisation (the SIYATH Foundation) took responsibility for monitoring the project. It has been heartening to see the positive effects brought about by these dedicated agents of change who helped the women to own up to responsibilities and make their own decisions. The farm was thriving as never before in 1988 when the severe ethnic violence and civil war that disrupted the country of Sri Lanka in the 1980s finally impeded the development of the women's farm. The ngo was unable to continue its support that was still necessary. In 1990, for the first time in its history, the members of the farm did not collectively cultivate the farm land. The majority of the women were very unhappy about this and wanted to find ways to keep their 'togetherness' and continue their enterprise. In 1993 the group had started again.

The last element of *positive perceptions of self which legitimise women's sense of dignity and self-respect* has been crucial in the history of the farm. The women learned new skills such as bookkeeping, new agricultural technologies, conducting group meetings, giving and receiving constructive criticism, analysing their own situation, arriving at collective decisions and also acting on these decisions. These processes contributed significantly to the women's self-esteem because they could prove to themselves and the outside world that they could do what nobody had expected poor women to do. With the increase in their self-confidence they were willing to take up more responsibilities regarding their own affairs.

According to the women themselves the most important change was that when the farm was thriving, none of its members needed to work as casual labourers for the local élite who used to exploit and humiliate them. This became obvious when the women began to stand up against exploitation and repression by local bureaucrats. No longer, for instance, did they agree to hand over their products free of charge to development officers. Nor did they refrain any longer from openly asking critical questions and voicing their own opinions. In view of the dominant gender and class relations, this was very unorthodox behaviour. However, it did help to serve their collective and individual interests much better.

This last element in women's autonomy cannot be considered separately from the other elements discussed above. The inter-relationship of the mental and the material aspects cannot be overemphasised. An improvement in economic output increases women's self-confidence and, at the same time, self-confidence is a prerequisite for poor women to stand up against those who try to keep them down.

CONCLUSION

The short history of the collective women's farm in Kurunduvila shows that poor women can acquire self-confidence to stand up for themselves and become more autonomous if there is the right kind of outside support in combination with a favourable political climate. This support should encourage them to collaborate instead of oppose each other and to make independent decisions on the basis of their own analysis of their situation. However, after ages of oppression it is not surprising that such a vulnerable process takes a very long time (Peiris and Risseeuw 1983; Risseeuw 1987).

The study also shows that women's autonomy, as an analytical concept, can be useful to direct and assess processes of change. Moreover, it involves the political ideal of creating a more equitable society in which women and men of different backgrounds would have equal opportunity and dignity (Schrijvers 1985: 234–7). At present this political objective appears to have mostly utopian connotations. The actual situation in which poor women in poor countries like Sri Lanka find themselves leaves little room for the realisation of autonomy for all. Macro economic and political factors that are out of reach at the local level have visibly widened the gap between the 'haves' and the 'have-nots' in Sri Lanka. The economic situation at rural grass-roots level has deteriorated. Over the past decade the country has been disrupted by violence and civil war which also affected the women's farm. Against this background, it is remarkable that the farm did not collapse totally when all outside support stopped and when some of the poorest members regressed to their former behavioural pattern of competing with each other.

Under these conditions of life and work, the concept of women's autonomy may appear over-idealistic or even irrelevant. However, precisely because of its explicitness regarding physical, economic, political and ideological space I prefer it in this context to the term 'empowerment'. Little is still known about the variations in gender relations and the scope for women's autonomy under different historical, cultural, class and ethnic conditions. South Asian feminist scholars and activists were the first to explicitly mention women's autonomy as a liberational condition in the process of transformation towards a better world (Jain 1983; Jayawardena 1982; 1986). In contrast to the concept of power or empowerment, autonomy is an anti-hierarchical concept which does not imply relations of domination. It expresses the principle of an attitude of inner strength, an attitude which makes room for transformation which comes from inner resources, moves bottom-up and works against undesirable domination. The importance of this mental aspect and the inter-relationship of the mental and material aspects of women's autonomy cannot be over emphasised. However, the question of whether the increased self-confidence of the members of the women's farm and their positive experience of doing and achieving things that nobody had expected of them will be strong enough to outlast a period of economic and political regression still remains. Outside support continues to be necessary. And above all, through my research, the extent to which micro changes at village level are interconnected with the wider political-economic structure is evident. Without structural transformation towards a more equitable and peaceful world at large, efforts to increase women's autonomy at the grass roots remain extremely vulnerable.

NOTE

1 The study formed part of a research project of the Research and Documentation Center on Women and Development, later Women and Autonomy (UENA), at Leiden University. The project was funded by The Netherlands Directorate General International Cooperation of the Ministry of Foreign Affairs.

10

WOMEN AND FAMILIES' ECONOMIC ORGANISATION IN A PUNJABI VILLAGE, PAKISTAN

Joyce Aschenbrenner

INTRODUCTION

The fieldwork on which the following study is based was conducted, for the most part, nearly twenty-five years ago, as research for my dissertation (1967). At that time, a feminist perspective was not encouraged in graduate studies and relatively little attention was paid to women's activities in most anthropological fieldwork. Notable exceptions were Eglar's (1960) work in Pakistan and Papanek's (1964) study of women in Islamic societies, among others. As an outside observer and female investigator in a society in which purdah was practised, I shared with these women the opportunity to interview and observe both men and women in a number of settings, including the privacy of the household. As a result of this opportunity which carried its own prescription to collect data which would not be available to a male investigator, I was able to see women in action in the Punjabi village in which I worked. While the material circumstances and even the social reality of such villages have, no doubt, altered in succeeding years, the data from my research remains useful as a case study of the conditions of women's roles and status. Thus, rather than representing an authoritative presentation of women's current status in Pakistan's village life, it is a reinterpretation of the conditions of the lives of women I came to know at that time in the context of subsequent scholarship on women.

The questions under consideration here are: what are the conditions under which women as a group exert influence in a society, in particular, what were the conditions of women's influence on events in the community I studied? Thus stated, the issue is one of relative (effective) power of men and women in particular situations rather than one of absolute power hierarchies. The assumption of absolute concepts of power, I would contend, is itself a male perspective; in my experience, women do not operate on the basis of such fixed ideas although they may give lip service to them. If we view Punjabi village women from a hierarchical social perspective, they would be considered to have little power to affect their life conditions, an assumption that is not shared here.

A view that is presented in this chapter – shared by Rogers (1978), Margery Wolf (1972) in her study of Taiwanese peasants, Aswad (1967) in her work on Arab women, and Riegelhaupt (1967) in looking at economic activities of Greek women – is that the power of women in society rests to a great extent on their participation in solidarity groups. Women's solidarity, then, depends on the following conditions:

a) women work together in groups;
b) they participate as groups in ritual and recreational activities;
c) they regularly share information and experiences; and,
d) they are able to maintain loyalties to other women.

In the following pages, I will look at these four points in the context of Faqiriawalla, the small Punjabi village in Pakistan where I worked during the period 1964–5. I will present and discuss observations on women's activism and influence I observed in the village as they relate to these conditions and will update my findings with some current information on village women in Pakistan.[1]

SHARED ECONOMIC ACTIVITIES

In joint households, mothers-in-law and daughters-in-law shared household tasks as well as work in the fields. Of the fifty farm households in the village, thirty were joint, consisting of parents and one or more married sons or brothers and their wives, along with unmarried or divorced daughters, who were farming together. I interviewed the women in twenty of these joint households, as well as in a blacksmith's and a shoemaker's households, asking about activities they participated in jointly. Of the twenty farming households, all the women worked together in the fields sharing wheat and other crops from harvest. In the Punjab, it is not uncommon for women to work in the fields which included planting, harvesting and cultivating. Rahman (1987) states that women in wealthier and in high-caste families generally do not work in the fields and a study in Lahore district notes that only poor women do so (Slocum *et al.* 1960: 311); however, in Faqiriawalla even women from the wealthy families did agricultural work.[2] The landowners in the village were of the Arain *jat*, who are traditionally of low status, yet some families were economically quite well off. Economic differences were reflected through power relations in the village politics and in respect by other villagers; apart from possession of modern agricultural equipment, differences between rich and poor families in terms of material possessions and lifestyles were not great.

In eight households, the women did all household tasks together including cooking. I was informed that when daughters-in-law quarrelled, the mother-in-law made the decision to have them cook meals separately. However, even then they continued to share many tasks. Typically, they took turns making *lassi* (a drink prepared from buttermilk) *ghee* (clarified butter) and dung cakes for fuel as well as carrying meals to the men in the fields. They washed clothes and utensils together at the wells. They also engaged in activities together and individually which resulted in cash sales including spinning, picking anise seed, and weaving drawstring for *shalwar* (loose trousers); sometimes they sold anise, even wheat or rice they harvested together and divided the money among themselves.[3]

There was a village perception that quarrels among daughters-in-law and mothers and daughters-in-law were the main cause of the break-up of joint households (see Mandelbaum 1988). Certainly, these relationships were often troublesome, but those between brothers and between fathers and sons were also frequently conflict-ridden. In my study of conflict in the village, I found that there were more serious conflicts among men related patrilineally than among women (1967). Sharma's findings concerning women in north-west India are similar; she reports that while women may quarrel more frequently than men, they make up more quickly because they are so interdependent (1980: 184).

There were a number of economic activities which women pursued together outside households. For example, women often met at public hearths (*sanja chullah*) in the evening to bake their bread together; the wives of *machis* or carriers kept the fires going and were paid with flour.[4] The same women functioned as midwives for which they were paid a few rupees. Much of the informal buying and selling was done by women so that women controlled a certain amount of cash for household purposes though larger transactions were handled by men.

Given the high degree of serious conflict in the village, the extent to which women did cooperate and work together was remarkable. This fact is in agreement with my subjective perception of considerable solidarity which generally exists among village women.

RITUAL AND RECREATION

The high point of village social life revolved around engagement and wedding celebrations and women were integrally associated with these activities. The bride was the central figure in marriage celebrations and her female relatives also had crucial roles in events relating to engagement and marriage.

Engagement negotiations involved visits back and forth between families, sometimes lasting as long as two years; the small gifts and meals were provided by the women of either family who also had a great deal of input into negotiations of dowry and other transactions between the families. For example, I recorded the following in my field journal:

We got into the middle of a marriage negotiation that was threatening to break down. . . . The girl's family which was very poor was asking for Rs 600 to be used for the girl's dowry, 20 *dholas* in gold ornaments, and 10 suits for the girl. The boy's aunt from his mother's side had gone to see the family and indicated agreement. However, village women had visited the boy's mother and remonstrated with her saying that was far too expensive, and the bridegroom's mother would go into debt that way. As a result, she had put pressure on her husband to break the engagement. While we were interviewing the husband, Ghulam Mohammad and Allah Joiaii (a woman) came in and a long discussion and argument over the matter began. Finally [Allah Joiaii and G. Mohammad] left the room for a while and consulted; when they came back they agreed that Rehmat should offer Rs 500, 6 *dholas* and the suits. It seemed agreeable to everyone there . . . I was rather surprised at the active role Allah Joiaii played in the proceedings.

The surprise I expressed here indicated my initial assumption that women did not play an openly active role in such transactions although I discovered that they often did. The fact that women interceded with the prospective groom's mother indicates that they functioned as a group to make their influence felt. While the marriage was initially negotiated through a male, the wife of a *nambardar* obviously played a critical role in the proceedings. In my interviews concerning recent engagements, I discovered that women were openly involved in about half of the arrangements while they undoubtedly worked behind the scenes in others. They often worked for marriage with their sisters' and brothers' children; nearly one-fifth of the recorded marriages that took place over a twenty-year period were between cousins on the maternal side. Sharma (1980) also found that women were influential in marriage arrangements in the villages of east Punjab.

In celebrations centring around engagements, women were the principal actors. I attended two engagement parties in which women, young and old, performed vigorous dances and sang songs with many verses which they had memorised over the years. One of these events coincided with the celebration of *Id* (prophet's birthday) and was preceded by a house-to-house invitation by a group of little girls who had painted their palms with *mehndi* (henna), singing and clapping hands as they progressed in a procession through the village.

At weddings, women dressed and cared for the bride; they sang songs and practised various rituals such as blocking the door by the bride's sisters when the groom's party arrived and passing a bowl of milk over the groom's head by the bride's mother. They distributed sweets on this and all occasions surrounding marriage. Later, when children were born, mothers and sisters were very much in evidence in caring for the new mother and child and

227

giving gifts to the new born as well as distributing sweets. Eglar also notes that in the Pakistani village in which she worked:

> the transactions involving actual gift exchange in which men take part are limited in number and take place only on a few important occasions . . . [while] the exchange of clothes, food, sweets, and the money given at times of sickness . . . and on other occasions – all these transactions are the sole responsibility of women.

(Eglar 1960: 141)

The women regarded these small transaction with in-laws very seriously, both as tokens of their good will and good faith in negotiations and as economic exchanges. One women told me that if an engagement was broken, all gifts had to be returned including the smallest such as a needle and thread. A woman would frequently cite a list of what she had given to a prospective or actual daughter-in-law in order to justify actions or defend herself against later accusations. Daughters-in-law in turn, sometimes complained that they did not receive a fair share of wheat or other supplies from a mother-in-law as a justification for visiting their family, and receiving gifts from relatives. These rationalisations of behaviour reveal the underlying significance of such exchanges in establishing and confirming social bonds. Because women were responsible for most of these transactions, they held an important key to social harmony or discord.

Formal religion was male oriented. Muslim ritual is dominated by men who pray together in the mosques while women pray at home. Still, the family celebrations that accompanied religious observances gave women a great deal of scope for social participation with other women. For example, on *Id*, women prepared the special sweets dishes for the holiday and visited other households; also, groups of little girls could be seen dressed in new clothes stitched by their mothers, playing, singing and dancing in the fields.

Women were involved in magico/religious rituals at births (e.g., naming ceremonies) and weddings that reflected 'popular' or indigenous beliefs; I observed a pre-wedding ritual in which a group of women held a ceremony (at which they asked a priest to officiate) in which an evil spirit was made to 'depart' from a nervous bride. A 1984 study *Women in Islam* notes that widespread participation in 'informal' religion and belief in the magical/divine powers of saints among women provides 'personalised spiritual consolation and emotional comfort . . . indigenous medical care (both physical and psychological) as well as colour and recreation through regularly-held ceremonies, fairs and feasts at shrines' (Khuhro 1984: 5). I attended a fair for a local *pir* or saint at a distant village with women from Faqiriawalla. Mandelbaum (1988) mentions the particular involvement of Muslim women in rites commemorating local saints or *pirs*.

In addition, the women of Faqiriawalla were very conscious about observance of prayer (*namaz*) and fasting during *Ramzan* (holy month). At that time, they kept religious differences with me as well as with my assistant (who was a Shiite) very much in the foreground. They were a force for religious conservatism in the village, serving to keep the male leadership mindful of its moral duties.[5]

Women's solidarity was expressed in daily routines as they groomed and gave physical care to one another, massaging and dressing skin and hair with oil. They expressed concern because I did not use oil on my hair (a criticism which as it turned out was well founded, since my hair did get very dry in the climate). A custom in the village when a family was robbed by a housebreaker illustrated female solidarity graphically. Female relatives from other villages came to support and mourn with the householders. In crises and on special occasions of all kinds, female relatives in particular, and to some extent neighbours, supported other women through customs and rituals.

SHARED INFORMATION AND EXPERIENCES

Women played a crucial role in the village informational network. Most of my information about village sexual mores and behaviour, marriage practices and domestic relations was provided by a *nambardar*'s wife and by other village women. Men also talked about matters with each other, but they were limited by not being free to talk about women's activities because of purdah restrictions. On the other hand, women talked freely about both men's and women's behaviour. Rogers states:

> Female solidarity, expressed in informal women's groups held together by a well-developed interhousehold female communication network, is most often cited as the strongest power base from which women operate in the community . . . [acting] as a kind of information control, heavily influencing community public opinion . . . their gossip also shapes public opinion, indirectly affecting male political decisions and behavior.

> (Rogers 1975: 735–6)

Knowledge that women could talk freely about their behaviour was certainly a restraining influence on male leaders in the village who often used tenets of Islam to justify their leadership role and to persuade others to agree with them. The effects of exchange of personal information and public opinion could be seen in a number of incidents, including one cited earlier that took place during a marriage negotiation. In another instance, a wealthy man from the village who wanted to take a second wife could not do so because 'half the village was against it'. Since his wife was apparently popular among women in the village, their opinion undoubtedly weighed on her side and that of her daughter since the man in question wanted a son! In another case, public opinion pressurised a man into letting his divorced daughter marry a younger man of her own choice rather than to a wealthier older man (who presumably offered him money); I was informed that Allah Joiaii and other women were active forces in this decision. Finally, I recall an instance in which women in a household informed me that a labourer's wife had literally 'bewitched' a son with the help of an older woman. I heard gossip about this affair from other sources as well and eventually the labourer and his wife (who were originally from another village) were forced to move from the village. This was despite the fact that he was apparently a valued worker in an area where farm labour was in great demand and that he was employed by a *nambardar*, a powerful man in the village.

The influence of women in these contexts was obvious and rested on the strong and reliable lines of communication existing between them as they worked together at home, at the wells, at the village hearths and as they recreated and visited. While they were restricted in many ways, most women in the village did not observe strict purdah; so that while they spent most of their time working in the household, they were relatively free to move about.[6] They also expressed opinions on all aspects of village life including political leadership and economic affairs. Although they did not participate in the formalised political and legal structure in the village, yet these women were very much aware of issues and used informal strategies to influence decisions. In her work on Zapotec community, Chinas contends that women play non-formalised public roles that are extremely important in peasant societies because they 'operate in a social universe separate from men and they are privy to information which men either do not or may not circulate among themselves and [because] certain types of movement and courses of action are open only to women' (Chinas 1973: 109). Because the village men were bound by rules of purdah not to discuss or to be privy to women's affairs, an important part of their communication network was missing. This gave women an edge which they used to their mutual benefit when they perceived their basic rights to be at stake.

Much of the inter-dependence among the women was based on shared experiences and feelings. Although the village women were open and direct in expression of feelings among themselves, they were constrained and inhibited by prevailing custom in expressing their feelings to men. This inevitably resulted in quarrelling and conflict but there was also a great deal of empathy, which I experienced as an outsider who was accepted to a degree by some of the women. One or two of the women addressed me as 'sister' and Allah Joiaii informed me that they had to look after me like one of their family. As a result of these displays of friendship, when I left the village, I felt bereft, as if I had lost something very valuable.

MAINTAINING LOYALTIES

If solidarity and information networks are to be maintained, there must be mutual trust. In the Pakistani village, as in any society, female unity could be threatened by any number of experiences which women as a group encountered.

The most potentially disruptive experience for women's relationships was, of course, marriage. I found the sister-sister relationship to be the strongest bond between kins as it was neither troubled by the economic conflict and competition that existed among brothers nor inhibited by the lines of authority between generations. When sisters married, however, they left home, while brothers remained. Nevertheless, sisters maintained ties after marriage. In contrast to the strict Hindu practice of village and kin exogamy, many Pakistani marriages occur within the village and between first cousins. Although marriage between paternal uncle's son and daughter are preferred according to Islamic custom, I recorded a relatively high proportion of all recorded marriages to be between mother's sister's children and between brothers' and sisters' children (1967). Many women voiced a preference for marriages between sisters' children, since they felt the strong bonds between sisters would ensure harmonious family relationships. Sharma notes that even in India, where village exogamy is widely practised, sisters maintain close ties after marriage (Sharma 1980: 184).

An important stated reason for marrying kin was the greater possibility that women in a household would then be related so that daughters-in-law would work together more easily. In one household in which I attended an engagement party, a brother and sister were both marrying cousins – their paternal uncle's daughter and son respectively. It was a joint household and they had grown up and played together in the courtyard as children. One might then expect the women of this household to continue to be strongly unified.

Another mitigating factor in the disruption of female bonds with marriage is the fact that the conjugal tie – which in American society supersedes all other family ties – is not as central in joint households. In the village, strong marriage relationships were sometimes actively discouraged by mothers-in-law because they were seen as potentially disruptive to the joint household. Other family relationships often took precedence over a marital one; a woman reported that her mother tried to break up her marriage when trouble developed in her brother's marriage with her husband's sister. Such 'exchange' marriages were said to lead to conflict in the extended kin group. I found such marriages, i.e., between a woman and her brother's wife's brother, more common in a study of a Muslim community in India, which I studied later in 1969. There where patrilineal kin ties were weakened – partly, I contended, because of political forces stemming from a history of political control and autocratic rule in a Sikh-dominated state – I found that marriages with mother's and father's maternal relatives were more prevalent (23 per cent) than those with father's paternal relatives (less than 1 per cent; 14 per cent in Faqiriawalla) or even 'exchange marriages' (6.7 per cent); they were somewhat more frequent than were marriages with maternal kin in

Faqiriawalla (19.2 per cent) (Aschenbrenner 1969). This suggests that a balance of power in marriage negotiations may shift towards mother's or father's side, depending upon political or economic conditions.

When the importance of affinal links in forging political alliances and obtaining economic assistance is considered, the influence of women on marriage arrangements emerges as an important factor in their voice in village matters. In Faqiriawalla, relatives on the maternal side often served as peace makers when patrilineal kin fought over inheritance of land and wells.

Purdah restrictions and the separation of sexes also worked to strengthen bonds between women. Papanek observes that since in segregated circumstances competence in various tasks is judged by other women, they depend less on men for self-esteem (1973: 314). In Faqiriawalla, women were dependent on each other for companionship, help in various tasks and emotional support. Of course, under the circumstances there was much quarrelling and conflict among women. At times it appeared at the outset as if women were deeply divided and supported the male power structure and policies in the village; nevertheless, women did regularly support other women, even as they protected rights of husbands, brothers and sons.

DISCUSSION

At the beginning of this Chapter, I stated that this is not an authoritative presentation of women's current status in Pakistani village life, nor does it deal with absolute power relationships between women and men. Rather, it is a discussion of the conditions of women's influence on events in the community at the time I studied it. In the social system of which the Punjabi village I have been describing is a part, men were unquestionably dominant in the formal economic and political structure. Pakistani laws, strongly influenced by Islam, favour men in marriage and in inheritance. Men can legally take four wives and they can divorce by merely pronouncing their intention in front of witnesses while women must go to court and prove desertion or abuse.[7] In the village, sometimes a woman forced a man to divorce her by going to her parents' house and refusing to function as his wife. However, even in their own families, daughters could inherit only one-half of a son's share. In actuality, they often gave up their inheritance to avoid a disruption of close ties to their natal family. Khuhro (1984) reports this situation and attributes it to ignorance of Islamic traditions among women; however, in my experience, village women knew about their inheritance rights. Mandelbaum (1988) and Sharma (1980) also describe this reluctance on the part of daughters to claim inheritance rights; the latter states:

> for a sister the good will of her brothers is important to her after her parents' death, and a woman will prefer to be in a position to be able to

call on her brothers for help of any kind, should she need to, after her marriage. This right may be of more advantage to her than the material gains she might obtain from inheriting a share in her father's land[8].

(Sharma 1980: 57)

This was the case in Faqiriawalla, also, indicating that within certain limits, women could manipulate the system to attain their ends in spite of discriminatory treatment. However, even individual strategies such as those used to obtain a divorce and to enlist the support of male kin were dependent on the active support and cooperation of female relatives. Men – brothers and fathers – were often prevented from giving active support to women even if they wished to, by an ideology that held women to be completely subservient to men with rights only through men. As women in the village worked, recreated, observed rituals together, shared feelings and knowledge and supported each other, they carved out an area of influence and authority which men could not undermine since their knowledge about women and their power to oppose them was limited because of separation of the domains of women and men.

Women seldom openly challenged men except when supporting the men of their families against others. Yet, jointly, they exerted influence on many of the decisions that were made. It was only when I realised the influence these women had that I could appreciate their willingness to express their views on most subjects forcefully as well as their visibility in village matters. On the basis of these observations, I would conclude that in affairs relating to marriage and property – the two most weighty concerns in the village – the women's collective views and interests were, on the whole, fairly well represented in Faqiriawalla.

When I first analysed the data from my research and wrote my ethnography, I was not able to articulate many of these ideas. I was at a loss in communicating the belief I held that women in the village had more power than the existing official system recognised. The many abuses of women's worth and dignity were apparent and duly recorded; yet, I felt this did not give the total picture of the role of women in village life.

If I were to do fieldwork in a similar social milieu today, I would design my research very differently. I would take greater note of the differences between women and men in responding to questions. I did record such differences, but I did not follow up the findings by attempting to find patterns and causes of the differences. Here, Rogers' (1978) and Ardener's (1982) observations about differences in women's and men's world views are appropriate. In my case, Rogers' claim that the neglect of such differences stems from a cultural bias on the part of anthropologists is well taken (Rogers 1978: 131). I had accepted the assumption of sameness in values and goals of men and women during my earlier academic endeavours, as a condition of my acceptance as an intellectual. A more reasonable alternative, which I now espouse, is that women's and men's views are identical when they have the same interests, but since women as a group often have interests different from men, their world views are often likely to diverge. When women primarily talk to other women about these interests, such differences may be overlooked by male investigators or by women whose work must be seen as acceptable by male mentors.

In the present study, it seems apparent that the solidarity of groups of women, combined with their economic importance in the village, gave them power and influence under certain conditions. These conditions pertained to family relationships, i.e, the need for cooperation in order to successfully farm and to be able to count on kin for defence or social support as well as to find a suitable mate for a son or daughter. In a society in which much of one's security derives from kinship, women, who are the primary keepers of kinship ties, have a natural advantage. When they build on this advantage by forming economically and socially interdependent groups, and by maintaining an active network of communication and sharing of information and opinion, their power as a group may be considerable. Here, I would take issue with Mandelbaum (1988: 19), who contends that family cohesion depends on women only in a negative sense in that quarrelling between women is a threat to family unity. Evidence presented here supports the view that village women actively worked to cement and maintain family ties.

The ideological status of village women is, of course, low; the birth of a son is a matter for rejoicing, that of girls is a matter for mourning: a woman in a family with newborn twin girls wailed, 'how can we poor people afford the dowries?'[9]

Current research on women indicates that Pakistani women are a relatively deprived segment of population in a male-dominated society and that in poor households, for example, they are discriminated against in terms of access to food, clothing and other needs (Qureshi 1983: 3). Programmes of development that were designed to provide social welfare, housing, family planning, education and health facilities would benefit women most. However, as in many societies in the world, military defence takes priority and such programmes have been limited. Further, as Weiss (1985) points out, economic planners often know very little about the economic contributions of women so that programmes of development often bypass them.

In, Faqiriawalla, the influence women exerted on affairs rested on family and economic matters that were, to some degree, independent of national policies regarding the rights of women, although such influence was certainly affected by national and international economics and by technological change. Thus, the tendency towards the Islamisation of the Pakistani legal system in the late 1970s and 1980s, the resulting protests and the establishment of a Women's Division in the Cabinet Secretariat (see Weiss 1985) exist on a level quite removed from the village women. Nevertheless, the Women's Division provides much-needed research on village women and has introduced a number of economic and social programmes for women at the village level (Nazeer and Aljarlaly 1983, 1984). As Singh and Kelles-Viitanen (1987) point out in their study of home-based production, the household cannot be treated as an entity separate from historical, cultural, socioeconomic and political contexts. These forces affected the village women in complex ways. While national political developments have influenced politics and law in Faqiriawalla, they have been greatly mediated by local custom and economy.

An investigation of the status of women in the Pakistani village today, then, would require the intensive kind of study that I conducted in 1964–5, emphasising household activities and family relationships placed in the context of national and international political and economic developments. The micro study of women's roles is particularly important in view of the dominance of the ideology of absolute power hierarchy in research on a wider scope, which may distort or obscure the role of women at the local level. It appears that the analysis of women's power in terms of absolute power hierarchies cannot account for anomalies in women's power as described here and that these ideologies in fact *produce* such apparent disparities because they obscure relative and situational power, as women create influence through their group activities and communication.

Programmes of improvement in peoples' lives should not be based merely on weaknesses, but need also to be built on the strengths displayed in their everyday lives. Thus, new programmes for women need to recognise and support their economic contributions as well as their solidary groups as they work, recreate and celebrate.

NOTES

1 The current data was courteously provided by Dr Sabeeha Hafeez, Director of Research, Women's Division, Government of Pakistan.

2 According to a study by the Research Wing of the Women's Division of the Government of Pakistan, as family income increases, the number of women who work in agricultural tasks decreases and that of men increases (1984: 4). However, the contributions of women at home-based and unpaid agricultural tasks remains uncounted in the national census. Nevertheless, in the village under study, it was observed that the economic contributions of women were recognised, imparting them a certain amount of influence in village affairs.

3 A study of eight villages in Gujarat District (Government of Pakistan Women's Division, 1984) indicated that women worked in the following household activities: childcare, cooking, washing pots, bringing firewood, carrying meals/tea/*lassi* to farms, knitting, sewing and weaving clothing, collecting animal waste, feeding, milking and caring for animals. A study of three villages in Peshawar district (Nazeer and Aljalaly 1983) listed women's farm activities as follows: poultry and animal husbandry, milking, vegetable picking, seed planting, shoveling, using sickles and carrying fuel and fodder. Other economic activities were teaching, nursing or midwifery, sewing, weaving, knitting/embroidery, tailoring, labour, petty business and match-making. The latter were paid activities although a

relatively small proportion of all village women were actually employed.

4 *Sanja chullah* is a hearth which is shared by women in the village. This system existed in rural Punjab and adjoining areas and provided women with a central meeting ground where they could bake *chappatis* (round flattened bread) and also exchange information, indulge in gossip and in general be themselves without the presence of their menfolk.

5 While common wisdom holds that women are more faithful about ritual observances, Khuhro notes that, 'practice of daily rituals is intermittent and irregular amongst rural women due to (a) poverty and the consequent hard struggle for existence and (b) ignorance'(1984: 2). Eglar (1960) found that Punjabi women in Mohla, Gujarat District, varied in degree of practice of *namaz* and other observances. My own observations are in agreement with this finding although women in Faqiriawalla were conscientious about fasting during *Ramzan*.

6 In general, observance of purdah is viewed as a status symbol (see, Rahman 1987; Mandelbaum 1988). In Faqiriawalla, women in wealthier families owned *burqas* (veils) wearing them on occasions when they left the village. Only the *Maulvi*'s (religious leader) wife and sister-in-law observed strict purdah, and seldom left their courtyard. Stringent observance of purdah was not advantageous as the women frequently needed to leave the household to work in the fields or at the wells.

7 According to the Muslim Family Laws Ordinance of 1961, a wife may obtain a divorce if the husband has neglected her maintenance for two years, if he takes a second wife without her permission, for desertion (a four-year period), if he is sentenced to imprisonment for seven years or more, has failed to perform marital obligations for three years, was impotent at the time of marriage and continues to be so, is insane or treats his wife with cruelty, or 'any other reason recognised as valid' (Government of Pakistan, Women's Division, 1981: 50–1).

8 A study of battered wives (Shamim *et al.* 1985) finds that most women whose husbands beat them belong to natal families that cannot or do not support them indicating that these concerns are realistic.

9 In the study of women in Gujarat District (Government of Pakistan 1984), 50 per cent of women interviewed agreed that dowry display was a curse and should be stopped or limited while 30–40 per cent (depending on the village) felt that the custom of dowry was in itself a bad practice.

Part IV
EPILOGUE

11

IN SUM
AND LOOKING BEYOND

Suggestions for future research S. Asia
                                    J16
Deipica Bagchi and Saraswati Raju   J21
                                           R23

IN SUM

The South Asian society is a complex mix of ethnicity, religion and culture. The three thousand and more years of Hindu culture have periodically assimilated influences of such diverse religions as Buddhism, Islam, Christianity and, in pockets, tribal animism. The original Hindu tenet towards women is derived from the principles enunciated by the ancient law giver Manu ordaining women to remain subservient to men accepting male patronage through various stages of life. Since caste evolved as the concept of racial purity, women have been viewed as potential bearers of caste pollution. Thus, the ideas of chastity, control over women's sexuality, restrictions on female mobility and autonomy – all amount to guarding the gateways for a racially pure society.

Over centuries of contact, the two dominant religion groups of the sub-continent have come to view women as 'preserver of the family honour'. While Islam brought about spatial and visual restrictions, it did not remain independent of the conservative elements of the land, which as some argue, neutralised whatever liberal ideology on women was contained in Islam. In some cases, the most conservative ideologies of the two religions seem to have joined hands in the subcontinent. Controls had to be lax on lower caste groups by virtue of economic necessities while restrictions became entrenched and rigid on the socially solvent group. What is also true is the phenomenon of 'Sanskritisation', or emulation by the lower caste (class) of the restrictive social norms of the upper-caste (class) Hindu society as the group moved up the economic ladder. There remained scattered groups, however, at the margins of the sub-continent such as along the remote valleys and mountains of the Himalayas or the islands off the peninsular coast that continue to demonstrate a social ethos somewhat untouched by the sub-continental gender dogma.

Our main objective in this collection has been to bring forth the regional variation in female labour participation at various levels trying to place

239

localised agroecological and/or techno-environmental, economic, demographic and ethnic/caste/religious contexts against the backdrop provided by broad regional and sub-regional tendencies. In the thematic organisation of our book, therefore, we adopted an analytical framework at different scales. Our contention is that such an approach not only enables one to test propositions back and forth at different levels, but also helps one resist the tendency of finding simplistic explanations for what appears on the surface. Once this stance is accepted, it is possible to understand the contradictions in the literature and resolve seemingly conflicting views regarding female labour.

The published data on female labour is dogged with problems of under-enumeration and conceptual flaws as well as inconsistent and inadequate measurement frameworks. Intriguingly, this invisibility in the data is actually a reflection on misplaced societal perception of female labour as of secondary consequence and the male as the primary bread-winner, despite conflicting evidence. However, within this essentially patriarchal vision which is reflected universally in conceptualising, reporting and recording of female contribution to the labour force, in South Asia, there exists considerable variation within and between countries. India provides a good case.

The recurring theme in terms of female labour has been the exclusion of household work in the official definition of work which places 'income' as the cut-off point between 'work' and 'non-work'. Empirical data contained in this volume make it possible to classify women's household work in cogent categories. In addition to paid work there are numerous time-consuming and labour-intensive activities that are both income generating and income saving such as provision of domestic resources which fall into the grey area of work. Similarly, time devoted to caring for the live-stock pertains to production of goods. The concept of 'chakra' as developed by Bagchi comes as close as is possible to defining the nature of this sector and to demonstating its continuum, motion, overlap, centrality and periphery.

More often, the reasoning that the household activities be viewed as economic is because while the female members are busy looking after hearth and home, other (male) members of the household may be freed to take up economically measurable productive activities. Given this, it has been increasingly argued that economic reappraisal of household activities is overdue.

Another argument to count the household work would be that these activities have a price tag, i.e., they can be substituted for goods and services which are otherwise paid for. However, this would entail inclusion of all except children, sick and elderly as workers with no temporal or spatial variation. As discussed in the introductory chapter, while this certainly would bring the crucial role of women to the forefront, it would also cloud the real issues, and a certain complacency may ensue because female labour would then no longer remain invisible. A more pertinent view, to which the editors subscribe, is to identify those aspects of household work which may fall outside the 'pure' domestic sphere as well as to expand the existing definition of 'work', to cover those activities which are not confined to the domestic sphere and yet are not considered as 'work'.

An amazing array of explanatory variables has been put forward to explain the regional variations in female participation in the labour force in South Asia, none of which appears to have a universal applicability in isolation. In general, the complex mix of ecological, economic, caste/class and gender-specific factors which are at work get mediated through space differently in different regions. Moreover, these constraints need to be interlinked with wider structures and processes in a society at large. Income and asset at household level, education and skill of females, dependency of population on earning members of the family and economic growth are intervening variables, but more often the region-specific absence or presence of social stigma associated with female work outside the familial domain assumes paramount importance.

In this context, not only the point that both rural as well as urban female labour, but also that the lower- and higher-caste females respond favourably to labour market conditions if a specific socioeconomic and historic environment becomes conducive to entry in the job market, is relevant. In general, the female labour market depends not only on good economic opportunities, but also on the kind of occupational avenues which are available for females. Given the biological and social construction of gender roles, women tend to concentrate in jobs where they can combine their household duties and 'work'. Thus, cottage and construction industries provide them with 'appropriate' chances. Interestingly enough, even the recent changes in the labour market, i.e. increasing contractual labour – essentially females and children working at home for the factory-based producers – seem to subscribe to the traditional order. The additional advantage is of course the fact that such labour is cheap and easily exploitable on various counts. This is true of all the countries across the sub-continent although the exact manner and degree to which this happens varies a great deal.

By and large, the prevailing gender ideologies seem to affect females irrespective of their traditional position in the caste/class hierarchy in terms of their participation in the workforce in the public sphere, albeit for different or rather overlapping reasons in the case of high-caste females. While the lower-caste females are constrained more on account of their relatively lower education and skill attainment rather than any restriction on their spatial mobility, the life worlds of Muslim females (Bangladesh and Pakistan) and those belonging to high castes (India) are limited further due to their particular caste and community membership.

In South Asia, there generally exists a parallel process of withdrawal of females from the labour market once the household's conditions improve.

This is essentially because of the misplaced patriarchal notion of the male as the primary breadwinner and also the notion of females as the custodians of the family's honour. Thus, restricting their spatial mobility becomes a way of controlling their sexuality outside the socially sanctioned bounds. However, this process may have different implications for females of different castes and groups in the sense that for a higher caste (and class) this may only mean a change of the scene from public to private space accompanied by an increased workload. Ironically, women from the lower socioeconomic order who earlier may have enjoyed a relatively higher degree of autonomy by virtue of their more active economic contributions to households get trapped into the restrictive lifestyles of the affluent and upper-caste (class) women.

In the overall context, however, it becomes crucial to analyse critically the processes at work resulting in females' withdrawal from the labour market from another angle. It has been argued that in the initial phases of the Green Revolution, demand for labour results in an increased labour force participation by females, but at a later stage they are pushed out and displaced by male labour and machines. However, our earlier discussion in the Introduction makes it clear that this relationship is not so simple and agricultural development may lead to casualisation of labour for both males and females in various degrees. What the available evidence suggests is that in some cases gender-specific explanations may be called for while in others gender relations may have to be placed in a wider structural context.

Notwithstanding the contention, it is true that within each category of class location at various scales, the female workers are posited at a disadvantageous and subordinate position *vis-à-vis* their male counterparts in terms of almost all aspects of work. This subordination cannot be captured in its entirety by the concept of class alone and regional context remains vital. Given all the complexities outlined earlier, following Walby (1990), we propose that the extended concept of patriarchy as a system of social structures seems to be central in our understanding of gender relations in the labour market. As argued by Walby, bringing social structure into the matrix helps us avoid the limitation of the classical patriarchal theory which places every individual man in the exploiting position and every women in a subordinate one (Walby 1990: 19–20). In our case, social structures can be constructed as the articulation (sum total) of a wide range of underlying constraints – explicit or subtle – mediated through space and expressed differently at different scales.

A logical outcome of our discussion so far is the asking of questions such as: do women enjoy no power at all in their respective caste–class and social context? Is the issue of power a relative phenomenon among men and women rather than being absolute in hierarchical terms? Is patriarchy invincible?

Time and again, the signals we receive from studies in this volume are that women's power relative to men rests to a great extent on their participation in solidarity groups and their capacity of networking. The rural females appear to be relatively better off than their urban counterparts in this respect with their extended kith and kin support. It has been observed that in extended rural households, across the sub-continent, women work together within and beyond their households and offer support and comfort to each other despite occasional differences. For social events such as marriages, betrothals, estrangements, deaths, etc., women are known to unite in remarkable solidarity to influence decision-making that would otherwise tilt in favour of the males. Of late, the most effective channels for change appear to be those that are set up by women themselves, drawing on their collective memories of shared lives. In South Asia, like most of the countries around the world, females through their collective action and memberships in solidarity groups and grass-roots organisations such as the well-known Self Employed Women's Association (SEWA), Working Women's Forum (WWF) in India, *Grameen* (rural) Bank in Bangladesh have been able to challenge the age-old conventions having a bearing on their lives.

While it is the traditional support system together with organised efforts that seem to have helped the rural females the most, in the urban context, access to education, skill formation and paid employment lending a sense of empowerment equip women to enter into a bargaining position with prevailing norms of patriarchy. This is a point we return to laconically in the end.

LOOKING BEYOND
SUGGESTIONS FOR FUTURE RESEARCH

I began to have an idea of my life, not as the slow shaping of achievement to fit my preconceived purposes, but as the gradual discovery and growth of a purpose which I did not know.

(Joanna Field (b.1900), English Psychologist)

At the risk of repetition it may be pointed out that a considerable amount of research has been devoted to establishing that conventional workforce statistics do not take into account certain activities when they are undertaken within the proximity of the home and are integrated with other domestic work. Since such activities are essentially carried out informally and do not always enter the commoditised market network which is the formal expression of economic activity, the workers engaged in these activities, primarily females, are not counted as workers, and they remain invisible. Although this 'invisibility' has generated much concern, from the practical point of view, the answer to the fundamental question as to 'what is work' still eludes the scholars. In the Indian context, those engaged in analyses of females' work are now arguing for the inclusion of time spent on food, fuel and water collection as work, and have started to question the classical capitalist/market-oriented definition of 'work'. However, precisely where the line between housework proper and the range of tasks within the household that constitute 'work' (to be acknowledged in the national counting system) should be drawn is not all that clear, and as Agarwal observes, the question continues to be part of an unresolved debate (1985: A159). As pointed out by Kalpagam (1986), from the academic point of view, 'invisibility' is an expression of marginalisation and non-integration of females into the main stream. In this context, analysis of and inquiry into wage formation, in so far as different aspects of housework can be considered as substitutable for wages, what Kalpagam calls, 'nexus between Family and Capitalism' under diverse sociocultural and economic realities assumes paramount importance.

Even as the housework lies outside the existing definition of work, the recently observed phenomenon of house-based contractual work which is intrinsically tied up with the formal market economy requires the traditional informal and formal dichotomy to be seen more as a continuum. But, as Duvvury observes, there is 'no rigorous analytical linking up' between the 'domestic economy [and] non-domestic social production' (1989a: WS 97). Here, what is extremely important is to distinguish between those females who voluntarily leave the household to train and work in the modern service sector, and those who are compelled to enter petty menial roadside jobs for mere survival of the family, what Bapat and Crook call the 'duality of female employment' (1988: 1591). These distinctions have ultimate bearing upon women's status in general and gender relations within the household in particular, and are essential for any proper assessment of the impact of development on women (Duvvury 1989a). However, in the existing literature, this duality has not received the attention it deserves.

Since the relationship between processes of development, including the shift from agriculture to non-agricultural contexts and absorption/exclusion of female workers, is not always unilateral in a positive or negative direction, time series data become crucial. For example, it is often argued that the complementary nature of the gender division of labour in the rural economy is replaced by a competitive one between units of labour in an urban situation resulting in a loss of employment to females, but we have virtually no studies where female (or, for that matter, male) migrants to cities are traced back to their rural roots to ascertain their work status prior to their move to the cities. Such a longitudinal approach to research on females is crucial in capturing the exact impact of urban migration on the shift of employment of women (Noponen 1990).

Some discussion on changes in migration patterns as they relate to the labour market becomes imperative and we propose to take it up subsequently in this section.

In the contemporary situations, one comes across what is termed as the 'fluidity of the gender division of tasks' (Saradamoni 1987), that is to say, the interchangability of tasks within a given job. In contrast, there exist cases where the traditional pattern of division of labour between the sexes has been upset. However, as pointed out by Duvvury (1989a: WS 108), we have very little information on the kind of stresses that may encourage this as well as break cultural barriers whereby women start pursuing new or forbidden avenues of employment. There is some information on the agricultural sector (Chatterji 1984; Saradamoni 1987; Mies 1988; Mencher in this volume), but information from non-agricultural urban situations is not forthcoming.

The invisibility as well as the marginality of women workers is sometimes justified in terms of the differential in productivity between male and female workers. However, as pointed out by Agarwal, efficiency measurement is problematic in the wake of sex-typing of tasks because when women are dominating in certain tasks, it is not possible to have a male/female comparative picture. Even where such studies are undertaken, the results have often been ambiguous. This is the area where a systematic probe has yet to be made (Agarwal 1985). Although Agarwal's observations essentially relate to rural situations, the same may be said about the urban context.

To come back to the question of house-based workers, available official statistics on them are highly contradictory and grossly underestimated. This is also because women themselves find it difficult to delineate a separation between all the activities in which they are concurrently involved in the home. It is evident that even for accounting the 'formal' work in an informal setting such as the home, operational definitions of work have to be evolved (Singh and Kelles-Viitanen 1987: 14). A related inquiry pertains to the time allocation of women in different tasks which cuts into their availability for the formal economy as well as leisure time, and is a reflection of the sexual division of labour in society. However, large-scale time-budget studies cannot be realistically undertaken by the national statistical organisations (Agarwal 1985; Krishnamurthy 1985). This entails thorough micro-level studies to be conducted by individual scholars for different regions such as the ones by Jain and Chand (1982). In this regard, tools and research methods used for data collection and research become extremely important. For example, the problems of rural women workers can be understood only in the context in which they are placed, particularly the family and the community structure, organisational membership, the institutional support, and above all, the region-specific sociocultural ethos. The choice of a particular method will depend upon the purpose of the investigation. For example, if the question is that of the situations or problems confronted by women, a few case studies of women from different landholdings and lower-caste categories may become relevant (such as undertaken by Gulati 1978, 1981); if on the other hand, the purpose is to carry out a large-scale analysis of women in agriculture in terms of their role in agriculture, wage structure, gender division of labour, technological changes and their implications for women workers, etc., field surveys would be more feasible involving structured and open-ended interviews (e.g. Bhalla 1989); but, if the purpose is to diagnose specifically the felt needs of women in agriculture, participatory research methods may be in order (Muthayya 1988).

The existing research on female participation in agriculture has mainly been concerned with a) the trends in agricultural labour, b) technological change and its impact on female employment in agriculture, or/and c) formulation of an agroecological framework for explaining inter-regional variation in female agricultural labour. In assessing the impact of technological changes, class considerations have been incorporated to some extent. However, it is surprising that in the South Asian context, especially in India, where societal taboos, restrictions on specific tasks, mobility, etc. become extremely caste-specific, there is very limited information on this aspect. As argued by Duvvury, the caste considerations have to be analysed in combination with other factors (1989a: WS 108). In some isolated cases, where the caste/class hierarchy, the sociocultural factors and the question of variation in regional ethos (the last one only peripherally) have been touched upon, the three components seem to have been included independent of each other and there has been no attempt to evolve a research framework to weave them into a coherent whole. That is to say, a research framework which takes into account technological, agroecological and sociocultural factors combined together in order to assess which of these factors overrides others – where and why – is lacking, although the present volume has tried to fill this gap.

As pointed out by Monk (1988), in considering women's work, changes in that work over time, and regional variation therein, narrowly focused ecological and economic theories and those social explanations that take into account only caste differences have to be rejected. Instead, these considerations need to be carefully integrated with more complex cultural-historical and societal interpretations.

This invariably brings in the concept of gender, which has increasingly been used in geographical research on women, as a critical one. The idea is that instead of being immutably ingrained in biological differences and therefore given, the distinctions between masculine and feminine are socially and culturally constructed and thus subject to change. However, the social construction of gender whereby women's social roles are fixed, i.e., looking after food preparation, taking care of children, elderly and sick, etc., pose key constraints on their spatial behaviour, what the Swedish geographer, Hagerstrand calls, 'space–time constraints'. There are a few studies on the gender–space (spatial mobility) relationship (Kala 1976; Sen 1988), but they essentially pertain to rural India, and we do not have any studies from urban locales.

Admittedly, by no means exhaustive, we have tried to highlight certain gaps in the existing research. There are other complex issues related to women and work such as occupational health hazards, state and voluntary/ kin support system, income generation and decision-making, and their impact on gender relations.

There are two more contemporary issues pertaining to labour which warrant our attention: a) the recent trend of diversification of rural economy in South Asia which necessitates some discussion on largely unexplored dimensions of migration pattern, and b) the process of economic restructuring.

In India, females outnumber males in rural-to-rural migration. Rural-to-urban migration is slightly more favourable to males than females, but when it comes to large urban centres, the stream becomes increasingly male selective. Of late, there has been some change in this pattern and areas dominated by male-selective inmigration have been witnessing family migration. Other things being equal, the labour force participation among the migrant population differs from that of the non-migrant population depending upon their status: life-time migrants *vis-à-vis* recent migrants as well as permanent versus temporary.

It has been convincingly argued by some that those who migrate to urban areas are not necessarily the 'dregs of society' who are pushed from their place of origin. The observation is consequent upon a relatively higher levels of skill, educational and employment attainments among the migrant population as compared to non-migrants at the place of destination, more so for specific sub-sections of female migrants, and holds for both India and Pakistan (Shah 1984; Premi 1985, 1990). However, it is interesting to note that generally more females migrate short distances as compared to males, but with literacy and a certain level of education this pattern changes and they travel longer distances (Singh 1984; Premi 1985). In Pakistan also, education seems to have a positive bearing on female employment in all categories of migrant females irrespective of their marital status. The association has been found to be stronger for single, divorced and widowed women in the older age group. However limited, these associations imply a certain degree of empowerment for females so as to question some of the role models and gender ideology regarding spatial mobility. But so far we are confronted with indirect inferences only.

Admittedly, our analysis is limited because of readily available information, but the existing literature for South Asia shows that as against the lifetime female migrants in older age groups, the recent young female migrants participate to a much lesser degree in the labour market. This is largely because a substantial proportion of recent female migration is related to marriage migration and in initial years housewifing takes precedence over employment. Among other things, this places (stages of) life-cycle considerations at the helm about which our information is extremely meager.

Another facet which is crucial for our understanding of the changing societal process at work pertains to employment-linked autonomous migration among females. As compared to some Asian countries, a minuscule

proportion of female migrants in South Asia can be included in this category, but the changing social and economic environment is bound to
trigger an increase in their numbers in future. These autonomous female
migrants would perhaps belong to two extreme ends of the occupational
hierarchy: highly educated and skilled in non-conventional jobs, or the most
destitute segments of society. In either case, it would be interesting to
observe the employment patterns and the underlying factors for their entry
into the labour market, essentially in urban areas.

The predominance of females in the rural-to-rural migration has traditionally been explained in terms of village exogamy and patrilocal residence. It
has been pointed out, however, that these practices are not common among
several low castes where the decision to move depends more upon demand
for agricultural labourers. Moreover, the increasing cultivation of cash crops
such as sugarcane, tobacco, cotton and woodcutting makes a heavy demand
on labour, particularly female labour, which participates extensively in
various tasks associated with these crops (Rao 1978; Bremen 1979; Mitra *et
al.* 1980; Ali 1981; Reddy 1981; Mies 1984; Duvvury 1986).

The share of rural non-agricultural employment in total employment in
India has risen in recent years from about 14 per cent in 1977–8 to 17 per
cent in 1987–8. This performance becomes even more impressive when it is
realised that the share of rural areas in total employment is actually declining
meaning thereby that employment in the non-farm activities has been
growing at a much faster rate than the overall employment rate in the
economy, on an average by about five times the rate of all rural employment
which incidentally is well above the rate of growth of all urban employment
(Bhalla 1992: 6–7). This diversification has its own implication for the
retention of rural population and slowing down of the rural migration to
urban areas. From the viewpoint of absorption of labour, however, what is
more relevant to note is that among the rural non-farm sectors, construction
where female participation has traditionally been substantial is emerging as
one of the most important activities whereby the share of rural areas in
construction has reached almost 70 per cent (Bhalla 1992; also, Jain 1979).
Given these recent changes in the rural economy, the preponderance of
females in the rural-to-rural migration can no longer be fully explained as
associational or marriage linked in nature. However, research on this aspect
is virtually non-existent.

Of late the growing concern regarding female labour is one which arises
from the changing economic order in South Asia. In keeping with the
restructuring of the global economy furthered by the internal trade deficit
and balance of payment crisis, the Indian government introduced the New
Economic Policy in July 1991. The Sri Lankan government had introduced
the programme of economic liberalisation way back in 1977 (Osmani 1987).
This marks a shift away from a regulatory regime and an acceptance of
export-led growth as an important viable development strategy. In order to

survive amidst severe international competition, a cost-effective labour pool becomes of utmost importance. Evidence from other countries shows that under such circumstances, the demand for flexible female labour increases substantially. This is essentially because female labour is relatively cheap, hard to unionise and through contractual practices can be hired to work at their homes or adjoining industrial premises. By doing so and by keeping the size of individual enterprises small, the employers can easily escape protective labour legislations of 'formal' settings and can dispense with female labour when their services are no longer required. What are the implications of this? Some have argued that in a situation where mere survival is at stake, demand for female labour, even if it is for lower skill, lower paid and lower occupational ladder (than males) in an unorganised, unprotective and exploitative environment is a welcome change from a situation of gross unemployment for them, at least as a short-term measure (Deshpande 1992). On the flip side of it, however, it has been pointed out that in structural adjustment policies, marketed goods, services and subsistence cash production are prioritised to such an extent that women's unpaid labour in reproduction and maintenance of human resources is seen as elastic. This gets reflected in resource appropriation and allocation in the form of cuts in social expenditure and subsidies which may have direct bearing on many practical gender needs. The perceived elasticity of female labour, in terms of increasing self-production of food and adaptation through changes in purchasing habits and consumption patterns, is expected to act as a shock absorber (Elson 1991: 24–5; Moser 1991: 104–5).

Given the formative nature of economic restructuring in South Asia, particularly in India, it is rather difficult to comprehend fully its implications for labour absorption. It is also too early to predict a conclusive outcome of the shift for the female labour. However, one point can be made with certain conviction: economic liberalisations cannot be expected to overturn existing inequalities and social institutions easily. While immediate employment prospects for females may appear promising, a few crucial questions still remain: now that female labour is clearly income generating and efficiency oriented and can no longer be simultaneously interchanged with household chores, would women's entry into the labour market be the result of a substitution for other work for them, or would it be in addition to their role as housewives resulting in a further increase in their total labour time? Put differently, would the increased work burden be accompanied by a changed sexual division of labour within the household?

At least in the first phase, new jobs are likely to spring up more in and around large urban centres because of their locational advantages where families are increasingly becoming nucleated. What kind of institutional support and state intervention would be available to replace the traditional sources of support? What would be the precise benefits of personal income for the females? What would be the chances for skill upgradation and mobility for these female workers once they are in? Would their work status ultimately lead to their empowerment and enhanced status?

It has been argued that the success of economic liberalisation in the South Asian context cannot be solely dependent upon export-led growth, and the growth of the domestic market will have a significant role in making the 'market-friendly' economies successful. Since the countries in the region are predominantly rural, it is the rural market that is involved. As the Indian 'reform' is very recent, it is difficult to make any definite statements about the implication of this expansion for demand and supply of labour. The potential for rural markets to expand depends upon increase in rural income. Given the existing social ethos, increase in income may result in a decline in visible work for certain sections of rural females accompanied by increasing labour participation by a larger segment of poor females consequent upon skewed income distribution. This would also imply some attempt towards literacy and skill acquisition by these females. What implications would this have on the existing gender relations in the rural society?

Answers to these questions hinge upon the painful and slow processes of social transformation and a fundamental change in the underlying gender ideology in the highly structured patriarchal societies of South Asia. An oblique reference to a very recent phenomenon in Andhra Pradesh (India) is worth quoting. The state has been in the news for village women's powerful drive against liquor shops. Instead of backing a socially relevant movement in an already impoverished context, the state government attempted to censor adult literacy textbooks which according to them seem to have enlightened these women to question and threaten the existing patriarchal power structure in the villages and face the powerful liquor lobby squarely (Kapoor 1992)!

A recent phenomenon in the Indian situation has been women's entry into the so-called male domain of work. In this context, some interesting statistics are in order. The last three decades in India have witnessed a threefold increase in the enrolment of female students for higher education. In hitherto unexplored areas such as engineering there has been a 28 per cent growth during the 1980s – three times what it has been in the 1950s. The number of women in the commerce field has grown five times as compared to a decade ago (Jain *et al*. 1992). A similar influx of female students has been witnessed in architecture, management, dairy farming and even marine technology. In the recent past, the Indian military services inducted women in their army, air force and navy wings. While all this is quite promising, a more pertinent question to ask is: does this really mean a loosening up of the patriarchal hold, or are women allowed to enter these non-conventional avenues to an extent that does not threaten the very basis of patriarchal ideology? In addition, one needs to know if the recently opening venues for females are in fact consequent upon the unavailability of the male colleagues who are no longer willing to perform particular tasks. In other words, is it as encroachment upon the male domain, or as replacement that the recent changes in the labour market should be viewed? That the induction of females in the military is in non-combat, educational and legal fields – the 'soft' options in the hard lives, that the pilots are to fly transport planes and the female engine driver the goods train, that when an all-female crew flight took place, the passengers were not told about it until after the safe landing of the aircraft – are a few examples that would justify our proposition that the recent inroads into the stronghold of fundamental patriarchal ideology is still very much guided by the 'this far and no further' dictum.

This is, however, not being cynical in undermining some watershed changes in the societal attitude. Even as the South Asian females operate within a wider patriarchal structure, what is more important to take note of is that a beginning has been made whereby the females have been able to engineer a dent which could some day split open the path all the way through.

NOTE

1 The primary responsibility of writing this epilogue has been with Deipica Bagchi and Saraswati Raju respectively.

N A to end

REFERENCES AND
SELECT BIBLIOGRAPHY

Abbasi, M.B. (1980) *Socio-Economic Characteristics of Women in Sind*, Karachi: Sind Regional Plan Organisation, Economic Studies Centre.

Abdullah, T.A. and Zeidenstein, S.A. (1975) 'Socio-economic implications of introducing HYV rice production on rural women in Bangladesh', paper presented at the International Seminar on Socio-economic Implications of Introducing HYV in Bangladesh, Bangladesh Academy of Rural Development, Ford Foundation, Dacca.

—— (1982) *Village Women of Bangladesh: Prospects for Change* Oxford: Pergamon for ILO.

Acharya, M. (1979) *Statistical Profile of Nepalese Women: A Critical Review* (vol. I), Kathmandu: CEDA, Tribhuran University.

—— (1981) *The Maithili Women of Sirsia (The Status of Women in Nepal)*, 1 (vol. II), Kathmandu: CEDA, Tribhuran University.

—— (1984) *Rastria Bikash Ra Nepali Aimai – Euta Naya Sandarva* (in Nepali), Kathmandu: UNICEF.

—— (1987) 'Λ study of rural labor markets in Nepal', Ph.D. dissertation University of Wisconsin, Madison.

Acharya, M. and Bennett, L. (1981) *The Rural Women of Nepal: An Aggregate Analysis Summary of 8 Village Studies (The Status of Women in Nepal)*, 9 (Part-II), Kathmandu: CEDA, Tribhuvan University.

—— (1982) 'Women in the subsistence sector: economic participation and household decision-making in Nepal', World Bank staff working paper no. 526.

—— (1983) '*Women and the subsistence sector: economic particiaption and household decision-making in Nepal*', World Bank staff working paper no. 562.

Acharya, S. (1987) 'Female labour participation in rural Asia', paper presented for a UNO Conference on Household Strategies and Labour, Centre for Women in Development, Kathmandu, Nepal.

Acharya, S. and Mathrani, V. (1993) 'Women in the Indian labour force: a sectoral and regional analysis' in A. Sharma and S. Singh (eds) *Women and Work Changing Scenario in India*, Patna: Indian Society of Labour Economics: 40–57.

Acharya, S. and Panwalkar, V.G. (1988) 'The employment guarantee scheme', regional research paper, The Population Council, Bangkok.

Adnan, S. and Islam, R. (1977) 'Social change and rural women: possibilities of participation', Proceedings, Role of Women in Socioeconomic Development in Bangladesh, Bangladesh Economic Association, Dhaka.

Afzal, M. (1989) 'Female labour force in South Asia–some issues', in K.K. Pathak, P.S. Bhatia and M.K. Premi (eds) *Population Transition in South Asia*, New Delhi: Indian Association for the Study of Population.

252

Aga Khan Foundation (1987) 'Information Bulletin', December, Geneva.

Aga Khan Rural Support Programme (AKRSP), Women in Development (1987) *Consultancy and Internships Report*, no. 10, Gilgit.

Agarwal, B. (1983) 'Women's studies in Asia and the Pacific', Asian and Pacific Development Centre, Kuala Lumpur, Malaysia Programme on Women in Development, *Occasional Paper* no. 4.

—— (1984) 'Rural women and high yielding variety rice technology', *Economic and Political Weekly*, 19 (13): A39–52.

—— (1985) 'Work participation of rural women in the third world: some data and conceptual bias', *Economic and Political Weekly*, 20 (51–2): A 155–64.

—— (1986a) *Cold Hearths and Barron Slopes*, New Delhi: Allied Publishers.

—— (1986b) 'Women, poverty and agricultural growth in India', *Journal of Peasant Studies*, 13 (4): 165–220.

—— (1988) 'Who sows? Who reaps? Women and land rights in India', *Journal of Peasant Studies*, 15 (4): 531–81.

—— (1989) 'Rural women, poverty and natural resources sustenance, sustainability and struggle for change', *Economic and Political Weekly*, 24 (43): WS 46–78.

Aguiar, N. (1982) 'Women in the labor force in Latin America: a review of literature', paper presented at the Panel on Research on Women: Towards a World Perspective, International Sociological Association, 10th World Congress of Sociology, Mexico, August.

Ahluwalia, M.S. (1978) 'Rural poverty and agricultural performance in India', *The Journal of Development Studies*, 14 (3): 298–323.

Ahmad, I. (ed.) (1976) *Family, Kinship and Marriage Among Muslims*, New Delhi: Manohar.

—— (1985) 'Gender studies and Muslim women in India', paper presented at the National Seminar on Relevance of Sociology, Department of Sociology, Aligarh Muslim University, November.

Ahooja-Patel, K. (1986) 'Employment of women in Sri Lanka: the situation in Colombo', in R. Anker and C. Hein (eds) *Sex Inequalities in Urban Employment in the Third World*, London: Macmillan Press.

Aizaz, N. (1989) 'Development and the prospects of absorption of Muslim women in the formal and informal sectors of economy: the scenario from north-western India', paper submitted for presention at the Commonwealth Geography Bureau Workshop on Gender and Development, University of Newcastle, England, April.

Alamgir, S.F. (1977) *Profile of Bangladesh Women; Selected Aspects of Women's Roles and Status in Bangladesh*, Dhaka: US Agency for International Development (USAID).

Ali, N. (1981) 'Some aspects of migratory labor in Jammu and Kashmir, paper presented at the National Seminar on Migration Labour in India, National Labour Institute, New Delhi.

Ambannavar, J.P. (1975) 'Changes in economic activity of males and females in India 1911–61', *Demography*, 4: 344–64.

Amin, A.T.M.N. (1986) 'Urban employment of underdevelopment', in R. Islam and M. Muqtada (eds) Bangladesh: Selected Issues in Employment and Development, New Delhi: ILO ARTEP.

—— (1987) *Statistical Pocket Book of Bangladesh*, Dhaka: Bangladesh.

Amjad, R. (ed.) (1989) *To the Gulf and Back: Studies on the Impact of Asian Labor Migration*, New Delhi: ILO/ARTEP.

Anandalakshmy, S. and Sawhney, H.K. (1988) 'Women's contributions to the productive process in agriculture: impact of modernization', paper presented at the

International Conference on Appropriate Agricultural Technologies for Farm Women, Indian Council of Agricultural Research in collaboration with International Rice Research Institute, Manila, Philippines, New Delhi.

Anker, R. (1983) 'Female labour force participation in developing countries: a critique of current definitions and data collection techniques', *International Labour Review*, 122 (6).

Anker R., Khan, M.E. and Gupta, R.B. (1987) 'Biases in measuring the labour force', *International Labour Review*, 126 (2): 151–67.

Ardener, S. (ed.) (1982) *Perceiving Women*, London: J.M. Dent.

Aschenbrenner, J. (1967) 'Endogamy and social status in a West Punjabi village', Ph.D dissertation, University of Minnesota.

—— (1969) 'Politics and Islamic marriage practices in the Indian subcontinent', *Anthropological Quarterly*, 42: 305–15.

ASTRA (1981) 'Rural energy consumption patterns: a field survey', Indian Institute of Science, Bangalore.

Asuri, T.P. and Mahadevappa, D. (1988) 'Women's participation in sericulture enterprises – a case study', paper presented at the International Conference on Appropriate Agricultural Technologies for Farm Women, Indian Council of Agricultural Research in collaboration with International Rice Research Institute, Manila, Philippines, New Delhi.

Aswad, B. (1967) 'Key and peripheral roles of noble women in a Middle Eastern Plains village', *Anthropological Quarterly*, 40: 139–52.

Bagchi, D. (1982) 'Female roles in agricultural modernization: a case of India', *WID Working Paper Series #10*, Office of Women in International Development, Michigan State University, East Lansing.

—— (1987) 'The domestic fuel crisis in India: impact on households and women', *WID Working Paper Series, #142*, Office of Women in International Development, Michigan State University, East Lansing.

Banerjee, N. (1985) *Women Workers in the Unorganized Sector*, Hyderabad: Sangam Books.

—— (1989a) 'Trends in women's employment, 1971–1981: Some macro-level observations', *Economic and Political Weekly* 24 (17): WS 10–22.

—— (1989b) 'Working women in colonial Bengal: modernisation and marginalisation', in K. Sangari and S. Vaid (eds) *Recasting Women Essays in Colonial History*, New Delhi: Kali for Women.

—— (1992) 'Poverty, work and gender in urban India', occasional paper no. 133, Centre for Studies in Social Sciences, Calcutta, India.

Bapat, M. and Crook, N. (1988) 'Duality of female employment – evidence from a study in Pune', *Economic and Political Weekly*, 23 (31): 1591–5.

Bardhan, K. (1984) 'Work patterns and social differention: rural women in West Bengal', in H.P. Binswanger and M.R. Rosenzweig (eds) *Contractual Arrangements, Employment and Wages in Rural Labor Markets in Asia*, Economic Growth Center Publication Series, 2207–17, Haven: Yale University Press.

—— (1985) 'Women's work, welfare and status forces of tradition and change in India', *Economic and Political Weekly*, 20 (51–2): 2261–9.

—— (1987) 'Women workers in South Asia: employment problems and policies in the context of the poverty target groups approach', ILO/ARTEP working paper P.B. no. 643, New Delhi.

—— (forthcoming) 'Women's work in relation to family strategies in South and Southeast Asia', paper for the United Nations University project on Women and Family Strategies in South and Southeast Asia.

Bardhan, P.K. (1974) 'On the incidence of poverty in rural India in the sixties', in P.K. Bardhan and T.N. Srinivasan (eds) *Poverty and Income Distribution in India*, Calcutta: Calcutta Statistical Publishing Society, pp. 264–80.

—— (1978) 'Some employment and unemployment characteristics of rural women: an analysis of NSS data for West Bengal, 1972–73', *Economic and Political Weekly*, 13 (12): A21–6.

—— (1990) 'Sex disparity in child survival in rural India', in T.N. Srinivasan and P.K. Bardhan (eds) *Rural Poverty in South Asia*, Delhi: Oxford University Press.

Bashir, K.M. (1988) 'Review of employment and unemployment statistics', in *Proceedings of the Second Seminar on Social Statistics*, Central Statistical Organisation, New Delhi.

Begum, S. (1983) 'Women and technology: rice processing in Bangladesh', *Proceedings of Conference on Women in Rice Farming System*, London: Gower Publishing, Centre for Urban Studies.

Beneria, L. (1988) 'Conceptualizing the labour force: the underestimation of women's economic activities', in R.E. Pahl (ed.) *On Work Historical Comparative and Theoretical Approaches*, Oxford: Basil Blackwell.

Beneria, L. and Sen, G. (1988) 'Accumulation, reproduction and women's role in economic development: Boserup revisited', in R.E. Pahl (ed.) *On Work Historical Comparative and Theoretical Approaches*, Oxford: Basil Blackwell.

Bennett, L. (1979) *Tradition and Change in the Legal Status of Nepalese Women*. Kathmandu: CEDA, Tribhuvan University.

—— (1981) *The Parbatiya Women of Bakundol, The Status of Women in Nepal*, vol. 2, part 7, Centre Economic Development and Administration, Tribhuvan University, Kathmandu, Nepal.

—— (1989) 'Mapping women's work: their place in India's agricultural labour force', paper for World Bank's Women in Development Review and the Country Economic Memorandum for India.

Berger, H. 'Das Hunza-Burushaski', (unpublished dictionary) Heidelberg.

Bhalla, G.S. and Kundu, A. (1982) 'Small and intermediate towns in India's regional development' in O.P. Mathur (ed.) *Small Cities and National Development*, Nagoya, Japan: UNCRD.

Bhalla, S. (1989) 'Technological change and women workers: evidence from expansionary phase in Haryana agriculture', Economic and Political Weekly, 24 (43): WS 67–78.

—— (1992) 'Employment and work force diversification in rural India', paper presented at seminar, 'Understanding Independent India', School of Social Sciences, Jawaharlal Nehru University, New Delhi, India.

Bhattacharya, S. (1985) 'On the issue of underenumeration of women's work in the Indian data collection system', in D. Jain and N. Bannerji (eds) *Tyranny of the Household: Investigative Essays on Women's Work*, Delhi: Shakti Books.

Bhatty, Z. (1981) *The Economic Role and Status of Women in the Beedi Industry in Allahabad, India,*, ILO, Geneva: Verlag Breitenbach Publishers.

—— (1987) 'Economic contribution of women to the household budget: a case study of the Beedi industry', in A.M. Singh and A. Kelles – Viitanen (eds) *Invisible Hands: Women in Home-based Production*, New Delhi: Sage.

Biswas, A.K., Bauer, J.G. and Rele, J.R. (1989) Sectoral distribution of the work-force in India: trends and projections. Census of India, occasional paper no. 3.

Bjorkman, J. (1986) *The Changing Division of Labor in South Asia, Women and Men in Indian Society, Economy and Politics*, Maryland: Riverdale.

Blakie, Cameron, and Seddon (1980) *Nepal in Crisis: Growth and Stagnation at the Periphery*, Bombay; Oxford University Press.

Boserup, E. (1970) *Women's Role in Economic Development*, London: George Allen & Unwin.

Boulding, E. (1976) *The Underside of History: A View of Women Through Time*, Boulder, Colorado: Westview Press.

Breman, J. (1979) 'Seasonal migration and cooperative capitalism: crushing of cane and of labor by sugar factories of Bardoli', paper presented at the ADC-ICRISAT Conference on the Adjustment Mechanism of Rural Labour Markets in Developing Areas, Hyderabad, India.

Brohier, R.L. (1975) *Food and the People*, Colombo: Lake House Investments Ltd.

Brow, J. (1978) *Vedda Villages of Anuradhapura: The Historical Anthropology of a Community in Sri Lanka*, Seattle: University of Washington Press.

Brown, C. (1982) 'Home production for use in the market economy', in B. Thorne and M. Yalon (eds) *Rethinking the Family*, New York: Longman.

Burton, M.L. and Reitz, K. (1981) 'The plough, female contribution to agricultural subsistence and polygyny: a long linear analysis', *Behavioral Science Research*, December.

Burton, M.L. and White, D.R. (1984) 'Sexual division of labor in agriculture, *American Anthropologist*, 86: 568–83.

Cain, G.G., Khanam, S. and Nahar, S. (1979) 'Class patriarchy and the structure of women's work in Bangladesh', Center for Policy Studies working paper no. 43, New York: Population Council.

Cecelski, E. (1987) 'Energy and rural women's work: crisis, response and policy alternatives', *International Labour Review*, 126 (1): 41–64.

Centre for Monitoring Indian Economy (CMIE) (1987, 1988) *Basic Statistics Relating to the Indian Economy*, Vols. I and II, Bombay.

Chadha, G.K. (1991) 'On measuring employment and earnings for weaker sections in rural India', *The Indian Journal of Labour Economics*, 34 (1): 28–40.

Chakravarty, K. and Tiwari, G.C. (1977) 'Regional variation in women's employment: a case study of five villages in three Indian states', mimeo, Programme of Women's Studies, ICSSR, New Delhi.

Charlton, S.E.M. (1984) *Women in Third World Development*, Boulder and London: Westview Press.

Chatterji, R. (1984) 'Marginalisation and the induction of women into wage labour', World Employment Programme research working paper no. WEP 10/WP 32, Geneva: ILO.

Chaudhury, R.H. and Nilufar, R.A. (1980) 'Female status in Bangladesh', Bangladesh Institute of Development Studies, Dhaka, Bangladesh.

Chen, L.C. (1982) 'Where have the women gone?', *Economic and Political Weekly*, 17 (10): 365–72.

Chen, M. (1983) *A Quiet Revolution: Women in Transition in Rural Bangladesh*, Cambridge, USA: Schenkman Publishing Company.

—— (1986) 'Poverty, gender and work in Bangladesh', *Economic and Political Weekly*, 21 (5): 217–22.

—— (1989) 'Women's work in Indian agricultural by agro-ecological zones: meeting needs of landless and land-poor women', *Economic and Political Weekly*, 24 (43): WS 79–89.

Chinas, B. (1973) *Isthmus Zapotecs: Women's Roles in Cultural Context*, New York: Holt, Rinehart & Winston.

Das, V. (1976) 'Indian women: work, power and status', in B.R. Nanda (ed.) *Indian Women: From Purdah to Modernity*, Delhi: Vikas Publishing House.

Dasgupta, S. and Maiti, A.K. (1986) *The Rural Energy Crisis, Poverty and Women's Roles in Five Indian Villages*, Geneva: ILO.

Deere, C.D. and Leal, M.L. de (1982) *Women in Andean Agriculture*, Geneva: ILO.

Desai, M.M. (1975) 'Economic opportunities for women', *Social Welfare*, 22 (6–7): 25–8.

Desai, N. (1984) 'Women's studies and the social sciences: a report from India', *Women's Studies International*, 3: 2–6.

Desai, N. and Krishnaraj, M. (eds) (1987) *Women and Society in India*, n.p.: Ajanta Publications.

—— (eds) (1989) 'Women studies in India', *Economic and Political Weekly*, 24 (29): 676.

Deshpande, S. (1992) 'Structural adjustment and feminization', *The Indian Journal of Labour Economics*, 35 (4): 349–57.

Devadas, R.P., Sundaran, P. and Sithalakshmi, S. (1988) 'Appropriate home science technologies with their implications to farm women and farming', paper presented at the International Conference on Appropriate Agricultural Technologies for Farm Women, Indian Council of Agricultural Research in collaboration with International Rice Research Institute, Manila, Philippines, New Delhi.

Directorate of Agriculture, Madhya Pradesh (1982) *Agricultural Statistics*, Bhopal, India.

Dixon, R. (1978) *Rural Women at Work: Strategies for Development in South Asia*, Baltimore: Johns Hopkins University Press.

—— (1982a) 'Counting women in the agricultural labour force', *Population and Development Review*, 8 (3): 539–66.

—— (1982b) 'Mobilizing women for rural employment in South Asia: issues of class, caste, and patronage', *Economic Development and Cultural Change*, 30 (2): 373–90.

Dixon-Mueller, R. and Anker, R. (1988) 'Assessing women's economic contributions to development', Training in Population, Human Resources and Development Planning, paper no. 6, Geneva: ILO.

D'Souza, V.S. (1959) 'Implications of occupational prestige for employment policy in India', *Artha Vijnana*, 1: 233–47.

—— (1969) 'Changing socioeconomic conditions and employment of women in India', in M.K. Chaudhry (ed.) *Trends of Socio-Economic Change in India 1871–1961*, Simla: Indian Institute of Advanced Study.

Dubey, L. (1969) *Matriliny and Islam: Religion and Society in the Laccadives*, Delhi: National Publishing House.

Duvvury, N. (1986) 'Commercial capital and agrarian structure: a case study of the Guntur tobacco economy', *Economic and Political Weekly*, 21 (30): PE 46–57.

—— (1989a) 'Women in agriculture: a review of the Indian literature', *Economic and Political Weekly*, 24 (43): WS 96–112.

—— (1989b) 'Work participation of women in India: a study with special reference to female agricultural labourers, 1961 to 1981', in A.V. Jose (ed.) *Limited Options – Women Workers in Rural India*, New Delhi: ILO/ARTEP.

Duvvury, N. and Issac, T.M. (1989) 'Women in the labour force: a discussion of conceptual and operational biases with reference to the recent Indian estimates', paper presented at National Workshop on Visibility of Women in Statistics and Indicators: Changing Perspectives, SNDT Women's University, Bombay, July.

Eapen, M. (1992) 'Fertility and female labour force participation in India', *Economic and Political Weekly*, 27 (40): 2179–88.

Edgren, G. (1987) *The Growing Sector*, Geneva: ILO.

Eglar, Z. (1960) *A Punjabi Village in Pakistan*, New York: Columbia University Press.

Ellman, A.O., Ratnaweera, D.D.S., de Silva, K.M. and Wickramasinghe, G. (1976) *Land Settlement in Sri Lanka 1840–1975. A Review of the Major Writings on the Subject*, Colombo: Agrarian Research and Training Institute, Research Study Series 16.

Elson, D. (1991) 'Male bias in the development process: an overview', in D. Elson (ed.) *Male Bias in the Development Process*, Manchester and New York: Manchester University Press.

Enayet, F.S. (1979) 'Economic activities and employment opportunities: the urban situation' in UNICEF, Women's Development Programme (ed.) *Situation of Women in Bangladesh*, Dhaka, Bangladesh.

Epstein, S. (1986) 'Cracks in the wall: changing gender roles in South Asia' in J.W. Bjorkmen (ed.) *The Changing Division of Labor in South Asia*, Riverdale, MD: Riverdale Company Publishers.

Esterline, M.H. (1987) *They Changed Their Worlds: Nine Women of Asia*, Lanham, MD: University Press of America.

Gadgil, D.R. (1965) *Women in the Working Force in India*, Bombay: Asia Publishing House.

Garnsey, E. and Paukert, L. (1987) *Industrial Change and Women's Employment: Trends in the New International Division of Labor*, ILO Research Series no. 86, Geneva: ILO.

Gerard, R. (1977) 'Feasibility study of production/income generating activities for rural women in Bangladesh', Women Development Programme, UNICEF, Dhaka, Bangladesh.

Government of India (1974) *Towards Equality: Report of the Committee of the Status of Women in India*, Delhi: Government Printing Press.

Government of Pakistan (1962) *Census of Pakistan, 1961*, Bulletin no. 2, Karachi: Manager of Publications.

Government of Pakistan, Women's Division (1981) *Report of the National Conference of Muslim Women*, Islamabad.

Government of Pakistan, Women's Division (1984) *Data Base on Women in Agriculture and Rural Households*, Islamabad.

Grover, D. and Krishnappa (1985) 'Female participation in work: demographic, developmental and social dimensions', in G.P. Mishra (ed.), *Regional Structure of Developmental Growth in India*, New Delhi: Ashish Publishing House.

Gulati, L. (1975a) 'Female work participation: a study of inter-state differences', *Economic and Political Weekly*, 10 (1–2): 35–42.

—— (1975b) 'Occupational distribution of working women: an inter-state comparison', *Economic and Political Weekly*, 10 (43): 1692–703.

—— (1976) 'Unemployment among female agricultural labourers', *Economic and Political Weekly*, 11 (13): 31–9.

—— (1978) 'Profile of a female agricultural labourer', *Economic and Political Weekly*, 13 (12): A27–35.

—— (1979) *Profiles in Poverty: A Study of Five Poor Working Women*, Trivandrum: Centre for Development Studies.

—— (1981) *Profiles in Female Poverty*, New Delhi: Hindustan Publishing Corporation.

—— (1984) *Fisherwomen in the Kerala Coast: Demographic and Socio-Economic Impact of Fisheries Development Project*, Geneva: ILO.

—— (1993) 'Women in the unorganised sector with special reference to Kerala', in A. Sharma and S. Singh (eds) *Women and Work Changing Scenario in India*, Patna: Indian Society of Labour Economics, pp. 40–57.

Gunawardena, R.A.L.H. (1973) 'Irrigation and hydraulic society in early medieval Ceylon', *Past and Present*, 53: 3–27.

Hagerstrand, T. (1967) *Innovation Diffusion as a Spatial Process* (trans. A. Pred), Chicago: University of Chicago Press.

Halim, A. and Florence, E.M. (1983) 'Women labourers in rural Bangladesh: a socioeconomic analysis', Graduate Training Institute, Mymensingh: Bangladesh Agricultural University.

Hare, H. (1981) *Women, Demography and Development*, Canberra: The Australian National University.

Harriss, B. and Watson, E. (1987) 'The sex ratio in South Asia' in J.H. Momsen and J.G. Townsend (eds) *Geography and Gender in the Third World*, London: Hutchinson.

Hassan, I. (1984) 'Rural women double burden and male attitude', *Pakistan Times Overseas*, November 21.

Helbock, L. (1975) 'The changing status of women in Islamic Pakistan', mimeo, Islamabad: USAID.

Hettige, S.T. (1984) *Wealth, Power and Prestige. Emerging Patterns of Social Inequality in a Peasant Context*, Colombo: Ministry of Higher Education.

Heyzer, N. (ed.) (1985) *Missing Women: Development Planning in Asia and the Pacific*, Kuala Lumpur: Asian and Pacific Development Centre.

Hippel, G. and Fischer, W. (1987) 'Entwicklung gegen die Frauen. Zur Situation der Frau im landlichen Nord-West Pakistan', *Entwicklund und Zusammenarbeit*, 10, 87: 12–13.

Hirway, I. (1992) 'Stabilisation, structural adjustment and the rural poor', *The Indian Journal of Labour Economics*, 35 (3): 296–302.

Honigman, J.J. (1957) 'Women in West Pakistan', in S. Maron (ed.) *Pakistan: Society and Culture*, New Haven: Human Relations Area Files.

Horowitz, B. and Kishwar, M. (1984) 'Family life – the unequal deal', in M. Kishwar and R. Vanita (eds) *In Search of Answers*, London: Zed Books Ltd.

Hossain, M. (1986) 'Employment generation through cottage industries: potentials and constraints' in R. Islam and M. Muqtada (eds) *Bangladesh: Selected Issues in Employment and Development*, New Delhi: ILO/ARTEP.

—— (1987) 'Employment generation through cottage industries – potentials and constraints: the case of Bangladesh', in R. Islam (ed.) *Rural Industrialisation and Employment in Asia*, New Delhi: ILO.

Hughes, A.W. (1978) *Gazeteer of the Province of Sind*, London: George Bell & Sons.

Huq, J.A. (1979) 'Economic activities of women in Bangladesh: the Rural situation' in UNICEF, Women's Development Programme (ed.) *Situation of Women in Bangladesh*, Dhaka, Bangladesh.

International Labour Office (ILO) (1987) 'Linking energy with survival: a guide to energy, environment and women's work', Geneva, Switzerland.

International Labour Office and the Institute of Social Studies (1987) *The Rural Energy Crisis, Women's Work and Basic Needs, Proceedings of an International Workshop*, Hague, Geneva, Switzerland.

Irfan, M. (1981) 'Studies in population: labour force and migration', project report no. 5, Islamabad: Pakistan Institute of Development Economics.

Islam, R. (1986) 'Rural unemployment and underemployment: a review in Bangladesh' in R. Islam and M. Muqtada (eds) *Bangladesh: Selected Issues in Employment and Development*, New Delhi: ILO/ARTEP.

Islam, R. and Shrestha, R.P. (1987) 'Employment expansion through cottage industries in Nepal: potentials and constraints', in R. Islam and M. Muqtada (eds) *Bangladesh: Selected Issues in Employment and Development*, New Delhi: ILO/ARTEP.

Jacobson, D. (1970) 'Hidden faces: Hindu and Muslim purdah in a central Indian village', Columbia University, Ph.D. Dissertation. Available through University Microfilms.

—— (1975) 'Separate spheres: differential modernization in rural central India' in H.E. Ullrich (ed.) *Competition and Modernization in South Asia*, New Delhi: Abhinav Publications.

—— (1976) 'The veil of virtue: purdah and the Muslim family in the Bhopal region of central India', in I. Ahmad (ed.) *Family, Kinship, and Marriage among Muslims in India*, New Delhi: Manohar Book Service.

—— (1976–7) 'Indian women in processes of development', *Journal of International Affairs* (Special Issue on Women and Change in the Developing World), 30 (2): 211–42.

—— (1977) 'Purdah in India: life behind the veil', *National Geographic Magazine*, 152 (2): 270–86.

—— (1978) 'The chaste wife: cultural norm and individual experience', in S. Vatuk (ed.) *American Studies in the Anthropology of India*, New Delhi: American Institute of Indian Studies and Manohar Publications.

—— (1982) 'Purdah and the Hindu family in central India', in H. Papanek and G. Minault (eds) *Separate Worlds: Studies of Purdah in South Asia*, Columbia, Mo.: South Asia Books, and Delhi: Chanakya Publications.

—— (1988) *Women and Work in South Asia: An Audiovisual Presentation*, St Louis Park, Minn.: Glenhurst Publications. (An audiovisual presentation available as a still video or filmstrip, issued by Women and Development Issues in Three World Areas, funded by the USAID Development Education Program).

—— (1992) 'The women of north and central India: goddesses and wives', in D. Jacobson and S.S. Wadley (eds) *Women in India: Two Perspectives*, New Delhi: Manohar Book Service, and Columbia, Mo.: South Asia Books.

Jain, D. (1979) 'Impact on women workers–Maharashtra Employment Guarantee Scheme: a case study', presented for the ILO, Institute of Social Studies, New Delhi, India.

—— (1983) *Development as if Women mattered or can Women build a new Paradigm*, New Delhi: Institute of Social Studies Trust.

—— (1985) 'The household trap: report on a field survey of female activity patterns', in D. Jain and N. Bannerji (ed) *Tyranny of the Household: Investigative Essays on Women's Work*, Delhi: Shakti Books.

Jain, D. and Chand, M. (1982) 'Report on a time-allocation study – its methodological implications', paper presented at Technical Seminar on women's work and employment, Institute of Social Studies Trust, New Delhi, April.

Jain, M., Rattani, L. and Rai, S. (1992) 'The changing woman', *India Today*, July 15: 52–61.

Jayawardena, K. (1982) *Feminism and Nationalism in the Third World in the 19th and early 20th Century*, The Hague: Institute of Social Studies.

—— (1986) *Feminism and Nationalism in the Third World*, London: Zed Books, New Delhi: Kali for Women.

Jayaweera, S. (1987) 'Class and gender in education and employment', paper presented at the UNO Conference in Kathmandu on the project Women's Work and Family Strategies.

Jeffery, P. Jeffery, R. and Lyon, A. (1989) 'Taking dung-work seriously: women's work and rural development in north India', *Economic and Political Weekly*, April 29: WS 32–7.

Jettmar, K. (1977) 'Sozialer wandel am Karkorum-Highway', *Indo-Asia*, 19: 48–55.

Jose, A.V. (ed.) (1989) *Limited Options – Women Workers in Rural India*, New

Delhi: ILO/ARTEP.

Joshi, H. (1976) 'Prospects and case for employment of women in Indian cities', *Economic and Political Weekly*, 11 (13): 1303–8.

Kabir, K., Abel, A. and Chen, M. (1976) *Rural Women in Bangladesh: Exploding Some Myths*, Dhaka: Ford Foundation.

Kakwani, N. and Subbarao, K. (1989) 'Poverty and its alleviation in India', draft version of paper for the World Bank.

Kala, C.V. (1976) 'Female participation in farm work in Kerala', *Sociological Bulletin*, 25 (2): 185–206.

Kalia, S.L. (1961) 'Sanskritization and westernization', in T.B. Naik & M. Bhouraskar, (eds) *The Changing Tribes of Madhya Pradesh*. Bhopal, Madhya Pradesh, India Government Press.

Kalpagam, U. (1986) 'Gender in economics: the Indian experience', *Economic and Political Weekly*, 21 (43): WS 59–66.

Kandiyoti, D. (1988) 'Bargaining with patriarchy', *Gender and Society*, 2 (3): 274–90.

Kannappan, S. (1985) 'Urban employment and the labour market in developing nations', *Economic Development and Cultural Change*, 33 (4): 699–730.

Kapadia, K. (1992) 'Every blade of green: landless women labourers, production and reproduction in south India', *The Indian Journal of Labour Economics*, 35 (3): 266–76.

Kapoor, A. (1992) 'Official meddling setback to literacy', *Times of India*, 26 December.

Karlekar, M. (1979) 'Balmiki women in India', paper prepared for the Indian Council of Social Science Research, New Delhi.

—— (1985) 'Some trends in women's studies in India', paper presented at the National Seminar on Relevance of Sociology, Aligarh Muslim University, November.

Kasturi, L. (n.d.) 'South Indian domestic workers in New Delhi', paper prepared for the Indian Council of Social Science Research, New Delhi.

Kazi, S. (1989) 'Some measures of the status of women in the course of development in South Asia', in V. Kanesalingam (ed.) *Women in Development in South Asia*, Delhi: MacMillan India Ltd.

Kemp, S.F. (1986) 'How women's work is perceived: hunger or humiliation', in J.W. Bjorkman (ed.) *The Changing Division of Labor in South Asia*, Riverdale, MD: Riverdale Company Publishers.

Khan, M. and Choudhury, N. (1986) 'Trade industrialization and employment', in R. Islam and M. Muqtada (eds) *Bangladesh: Selected Issues in Employment and Development*, New Delhi: ILO/ARTEP.

Khan, N.S., Shaheed, F., Mitha, Y. and Rahman, S. (1989) 'Income generation for women: lessons from the field in Punjab Province, Pakistan', *South Asia Bulletin*, 9(1): 26–46.

Khan, S. (1988) *The Fifty Percent: Women in Development and Policy in Bangladesh*, Dhaka: University Press Ltd.

Khuda, B. (1982) 'The use of time and underemployment in rural Bangladesh', University of Dhaka, Bangladesh.

Khuhro, M. (1984) *Women in Islam*, Islamabad: Women's Division, Government of Pakistan.

Kishwar, M. (1987) 'Toiling without rights: Ho women of Singbhum', *Economic and Political Weekly*, 22 (3, 4, 5): 95–101, 149–55, 194–200.

Kreutzmann, H. (1986) 'A note on yak-keeping in Hunza (northern areas of Pakistan)', *Production Pastorale et Societe*, 19 automne 99–106.

—— (1987) 'Die talschaft Hunza (northern areas of Pakistan): wandel der austauschbeziehungen unter einflub des Karakoram Highway', *Die Erde*, 118: 37–53.

—— (1989) 'Hunza – Landliche Entwicklung im Karakorum', Ph.D. dissertation, Institut für geographische Wissenschaften Freie Universitat Berlin

Krishnamurthy J. (1985) 'The investigator, the respondent and the survey: the problems of getting good data on women', in D. Jain and N. Banerjee (eds) *Tyranny of the Household*, New Delhi: Shakti Books.

Krishnamurty, J. (1990) 'Unemployment in India: the broad magnitudes and characteristics', in T.N. Srinivasan and P.K. Bardhan (eds) *Rural Poverty in South Asia*, Delhi: Oxford University Press.

Krishnaraj, M. (1988a) 'Women's studies: case for a new paradigm', *Economic and Political Weekly*, 23 (18): 892–4.

—— (1988b) 'Perspectives for women's studies', *Economic and Political Weekly* 23 (33): 1707–8.

—— (1990) 'Women's work in Indian census: beginnings of change', *Economic and Political Weekly*, 25 (48–9): 2663–72.

Krishnaraj, M. and Ranadive, J. (1982) 'The rural female heads of household hidden from view', Bombay: Research Centre for Women's Studies.

Kumar, S.K. (1978) 'Role of the household economy in child nutrition at low incomes', Department of Agricultural Economics, Ithaca; New York: Cornell University, occasional paper no. 95.

Kumari, V.V.L. and Chari, M.S. (1988) 'Participation of women in FCV-tobacco cultivation and its implications for research and development', paper presented at the International Conference on Appropriate Agricultural Technologies for Farm Women, Indian Council of Agricultural Research in collaboration with International Rice Research Institute, Manila, Philippines, New Delhi.

Kundu, A. and Premi, M.K. (1992) 'Work and non-work in the official statistical system: issues concerning data base and research on women in India' in A. Bose and M.K. Premi (eds) *Population Transition in South Asia*, Delhi: B.R. Publishing Corporation.

Kundu, A. and Rao, J.M. (1986) 'Inequity in educational development: issues in measurement, changing structure and its sociological correlates with special reference to India', in M. Raza (ed.) *Educational Planning: A Long Term Perspective*, New Delhi: National Institute of Educational Planning and Administration.

Kundu, A. and Raza, M. (1982) *Indian Economy: The Regional Dimension,* New Delhi: Spektrum Publishers.

Lateef, S. (1986) 'Ethnicity in India: implications for women', in J.W. Bjorkman (ed.) *The Changing Division of Labor in South Asia*, MD Riverdale Company Publishers.

—— (1990) *Muslim Women in India – Political and Private Realities 1890s–1980s*, New Delhi: Kali for Women.

Leach, E.R. (1961) *Pul Eliya, a Village in Ceylon. A Study of Land Tenure and Kinship*, Cambridge: Cambridge University Press.

Lessinger, J. (1985) 'Caught between work and modesty – the dilemma of women traders of Madras', *Manushi*, 28.

Libbee, M. (1977) 'Geographic research on women: speculations on the use of data about women', paper presented at the meeting of Association of American Geographers, April.

Lorimer, E.O. (1939) *Language Hunting in the Karakoram*, London: George Allen & Unwin.

Lynch, O.M. (1969) *The Politics of Untouchability: Social Mobility and Social Change in a City of India*, New York: Columbia University Press.

Maclachlan, M.D. (1983) *Why They Did Not Starve: Biocultural Adaptation in a*

South Indian Village, Philadelphia: Institute for the Study of Human Issues.

Madgulkar, V. (1958) *The Village Had No Walls*, Bombay: Asia Publishing House.

Malathy, R. (1993) 'The demand for non-market time: the case of married women in an urban setting', in A. Sharma and S. Singh (eds) *Women and Work: Changing Scenario in India*, Patna: Indian Society of Labour Economics, pp. 40–57.

Mandelbaum, D.G. (1988) *Women's Seclusion and Men's Honor: Sex Roles in North India, Bangladesh, and Pakistan*, Tucson: University of Arizona Press.

Marum, E. (1981) *Women in Food for Work*, Washington DC: USAID.

Mehta, M. (1990) *Employment Income and the Urban Poor*, New Delhi: Harman Publishing House.

Mencher, J. (1978) *Agriculture and Social Structure in Tamil Nadu: Past Origins, Present Transformations, and Future Prospects*, New Delhi: Allied Publishers, and Durham, North Carolina: Carolina Academic Press.

—— (1980) 'The lessons and non-lessons of Kerala: agricultural labourers and poverty', *Economic and Political Weekly*, 15 (41–3): 1781–802.

—— (1982) 'Agricultural labourers and poverty', *Economic and Political Weekly*, 17, (1–2): 38–43.

—— (1985a) 'The forgotten ones: female landless labourers in southern India', in *Women Creating Wealth: Transforming Economic Development*, Washington, DC: JT & A, Inc. for the Association for Women's Development.

—— (1985b) 'Landless women agricultural labourers in India: some observations from Tamil Nadu, Kerala and West Bengal', in *Women in Rice Farming: Proceedings of a Conference on Rice Farming Systems*, Brookfield, Vermont: Gower Publishing for the International Rice Research Institute, Los Banos, Philippines.

—— (1988) 'Women's work and poverty: women's contribution to household maintenance in south India', in D. Dwyer and J. Bruce (eds) *A Home Divided: Women and Income in the Third World*, Stanford: Stanford University Press.

—— (forthcoming) 'South Indian female cultivators and laborers: who are they and what do they do?', Michigan State University, working papers on Women in International Development.

Mencher, J. and D'Amico, D. (1987) 'Kerala women as labourers and supervisors: implications for women and development', in L. Dube, E. Leacock and S. Ardener (eds) *Visibility and Power: Essays on Women in Society and Development*, New Delhi: Oxford University Press.

Mencher, J. and Saradamoni, K. (1982) 'Muddy feet, dirty hands: rice production and female agricultural labour', *Economic and Political Weekly,* 17 (52): A149–67.

Mies, M. (1981) 'Dynamics of sexual division of lace workers of Narasapur', *Economic and Political Weekly*, 16 (10–12): 487–500.

—— (1984) 'Indian women in subsistence and agricultural labour', WEP research working paper, ILO, Geneva.

—— (1988) 'Class struggles and women's struggles in rural India', in M. Mies, V. Bennholdt-Thomson and C.V. Werlhof (eds) *Women: The Last Colony*, New Delhi: Kali for Women.

Miller, B. (1981) *The Endangered Sex: Neglect of Female Children in Rural North India*, Ithaca, New York: Cornell University Press.

—— (1982) 'Female labour participation and female seclusion in rural India: a regional view', *Economic Development and Cultural Change*, 30 (4): 777–85.

—— (1989) 'Changing patterns of juvenile sex ratio in rural India, 1961 to 1971', *Economic and Political Weekly*, 24 (22): 1229–36.

Mines, M. (1975) 'Islamization and Muslim ethnicity in south India', *Man: Journal of Royal Anthropological Institute*, 10 (3).

Minge-Klevana, W. (1980) 'Does labor time decrease with industrialization: survey of time allocation studies', *Current Anthropology*, 21: 279–87.

Mishra, R. P. (ed.) (1978) *Million Cities of India*, New Delhi: Vikas Publishing House.

Mitra, A. (1979) *The Status of Women: Literacy and Employment*, New Delhi: Allied Publishers Pvt. Ltd.

Mitra, A. and Mukhopadhyay, S. (1989) 'Female Labour in the construction sector', *Economic and Political Weekly*, 24 (10): 523–8.

Mitra, A., Mukherji S., and Bose, R.N. (1988a) *Indian Cities, their Industrial Structure, Inmigration and Capital Investment 1961–71*, New Delhi: ICSSR/JNU Publication.

Mitra, A., Mukherji, S., Bose, R. and Ray, L. (1981) *Functional Classification of India's Urban Areas by Factor Cluster Method 1961–1971* New Delhi: Abhinav Publlications.

Mitra, A., Pathak, L.P. and Mukherjee, S. (1980) *The Status of Women: Shifts in Occupational Participation 1961–71*, New Delhi: Abhinav Publications.

Mitra, A., Srimany, A.K. and Pathak, L.P. (1979) *The Status of Women: Household and Non-household Economic Activity*, Bombay: Allied Publishers.

Mitra, B.N., Lahiri, D. and Mahapatra, S.C. (1988b) 'Participation of women in rice-based farming system and its implications in overall development of agriculture – a case study', paper presented at the International Conference on Appropriate Agricultural Technologies for Farm Women, Indian Council of Agricultural Research in collaboration with International Rice Research Institute, Manila, Philippines, New Delhi.

Mohiuddin, Y.A. (1980) 'Women in urban labour market', *Pakistan Economist*, April 5: 24–9.

Molnar, A. (1981) *The Kham Magar Women of Thabang (The Status of Women in Nepal)*, Kathmandu: CEDA, Tribhuvan University.

Momsen, J. and Townsend, J. (ed.) (1987) *Geography of Gender in the Third World*, London: Hutchinson.

Monk, J. (1988) 'Encompassing gender: progress and challenges in geographic research', address given at the International Geographic Union meeting, Sydney, Australia, September.

Monk, J. and Hanson, S. (1982) 'On not excluding half of the human in human geography', *The Professional Geographer*, 34 (1): 11–23.

Moser, C.O.N. (1991) 'Gender planning in the Third World: meeting practical and strategic needs', in R. Grant and K. Newland (eds) *Gender and International Relations*, Milton Keynes: Open University Press.

Mujeeb, M. (1972) *Islamic Influence on Indian Society*, Meerut: Meenakshi Prakashan.

Mukerjee, A.B. (1971) 'Female participation in agricultural labour in Uttar Pradesh – spatial variations', *National Geographer* 6 (6): 13–18.

Müller-Stellrecht, I. (1979) *Materialien zur Ethnographie von Dardistan (Pakistan). Part i. Hunza*, Graz: Akademische Druck- und Verlagsanstalt.

Mumtaz, K. and Shaheed, F. (1970) (eds) *Women of Pakistan*, London: Zed Books.

Muthayya, B.C. (1988) 'Method of social research for socioeconomic studies on women in agriculture', paper presented at the International Conference on Appropriate Agricultural Technologies for farm women, Indian Council of Agricultural Research in collaboration with International Rice Research Institute, Manila, Philippines, New Delhi.

Nagaraj, K. (1989) 'Female workers in rural Tamil Nadu – a preliminary study', in A.V. Jose (ed.) *Limited Options – Women Workers in Rural India*, New Delhi: ILO/ARTEP.

Nath, K. (1965) 'Women workers in the new village', *Economic Weekly*, 17: 813–16.
—— (1970) 'Female work participation and economic development: a regional analysis', *Economic and Political Weekly*, 5 (21): 846–9.
—— (1975) 'Women in agriculture', *Indian Farming*, 25 (8): 5–71.
National Commission on Self Employed Women and Women in the Informal Sector (1988) *Shramshakti*, New Delhi: Government of India.
National Institute of Urban Affairs (NIUA) (1987) 'Women vendors in India's urban centres', Research Study Series no. 22, New Delhi: NIUA.
—— (1990) *Role of Women in the Urban Informal Sector,* New Delhi: NIUA.
—— (1991) 'Women in the urban informal sector', Research Study Series no. 49, New Delhi: NIUA.
National Sample Survey Organisation (NSSO) (1985) *Sarvekshana*, 8 (3–4); 11 (4).
—— (1986) *Sarvekshana*, 8 (4).
—— (1988) 'Concepts and procedures adopted in employment and unemployment surveys of the NSSO – a critical evaluation and suggestions for improvement', in *Proceedings of the Second Seminar on Social Statistics*, New Delhi: CSO, Government of India.
Nayyar, R. (1987) 'Female participation rates in rural India', *Economic and Political Weekly*, 23 (51): 2207–16.
—— (1989) 'Rural labour markets and employment of women in Punjab-Haryana', in A.V. Jose (ed.) *Limited Options – Women Workers in Rural India*, New Delhi: ILO/ARTEP.
Nazeer, M.M. and Aljalaly, S.Z. (1983) *Participation of Women in Cottage and Small Scale Industries*, Islamabad: Women's Division, Government of Pakistan.
Nepal Rastra Bank (1988) *Multi-Purpose Household Budget Survey, 1984–85*, Kathmandu: Nepal Rastria Bank.
Newland, K. (1980) 'Women, men and the division of labour', Worldwatch Paper 37, Washington DC.
Noponen, H. (1990) 'Surveying poor women and households over time: a family labor supply and survival matrix', Center for Urban Policy Research, working paper no. 2, Rutgers University, New Brunswick, USA.
Office of the Registrar General and Census Commissioner, (1961, 1971) Census of India, General Population Tables, Part II-A, Ministry of Home Affairs Government of India, New Delhi.
—— (n.d.) Census of India, 1981, 'Instructions to enumerators for filling up the household schedule and individual slip', mimeo.
—— (1991a) Draft tables with notes, Ministry of Home Affairs, Government of India, New Delhi.
—— (1991b) 'Instructions to enumerators for filling up the household schedule and individual slip', mimeo.
—— (1991c) Census of India, Paper 3 of 1991, Ministry of Home Affairs, Government of India, New Delhi.
Omvedt, G. (1980) 'Adivasis, culture and modes of production in India', *Bulletin of Concerned Asian Scholars*, 10: 15–22.
Osmani, S.R. (1987) 'The impact of economic liberalisation on the small-scale and rural industries of Sri Lanka', in R. Islam (ed.) *Rural Industrialisation and Employment in Asia*, New Delhi: ILO.
Panikar, P.G.K. (1983) 'Adoption of high yielding varieties of rice in Kerala: a study of selected villages in Palghat and Kuttanad', paper, Centre for Development Studies, Trivendrum, India.
Papanek, H. (1964) 'The women field workers in a purdah society', *Human Organization*, 23: 160–3.

—— (1973) 'Purdah, separate worlds and symbolic shelter', *Comparative Studies in Society and History*, 15: 285–325.

—— (1975) 'Women in south and south-east Asia: issues and research', *Sign: Journal of Women in Culture and Society*, 1 (1): 193–213.

—— (1989) 'Family and status-production work: women's contribution to social mobility and class differentiation', in M. Krishnaraj and K. Chanana (eds) *Gender and the Household Domain*, New Delhi: Sage.

Papola, T.S. (1986) 'Women workers in the formal sector of Lucknow, India', in R. Anker and C. Hein (eds) *Sex Inequalities in Urban Employment in the Third World*, London: Macmillan.

—— (1992) 'Rural non-farm employment: an assessment of recent trends', *The Indian Journal of Labour Economics*, 35 (3): 238–45.

Pariwala, R. (1985) 'Discussant's remark in section on "women and work" ', in K. Saradamoni (ed.) *Women, Work and Society*, Calcutta: Indian Statistical Institute.

Parthasarathy, G. and Ramarao, G.D. (1974) *Employment and Unemployment of Rural Labour and the Crash Programme*, Waltaire; Andhra University Press.

Pastner, C.M. (1973) 'Social dichotomization in society and culture: the women of Panjgur, Baluchistan', Ph.D. dissertation, Brandeis University.

—— (1974) 'Accommodations to purdah: the female perspective', *Journal of Marriage and the Family*, 36 (2): 36–42.

Patel, V. (1989) 'Recent debate on concepts of paid–unpaid work, domestic work and employment–unemployment of women', background paper submitted to National Workshop on Visibility of Women in Statistics and Indicators: Changing Perspective, SNDT University, Bombay, July.

Peake, L. (1989a) 'Women's work, incorporating gender into theories of urban social change', *Women and Environment*, 11 (3–4): 9–11.

—— (ed.) (1989b) 'Arena', *Journal of Geography in Higher Education*, 13 (1): 85–121.

Peiris, K. and Risseeuw, C. (1983) *It is your own Hand which gives you Shade: a Study on the Potential of Small Scale Coir Organisations of Women Workers in Sri Lanka*, Utrecht: Consultants for Management of Development Programmes.

Peries, O.S. (ed.) (1967) *The Development of Agriculture in the Dry Zone*, proceedings of a Symposium of the Ceylon Association for the Advancement of Science, Colombo: Swabhasha Printers.

Postel, E. and Schrijvers, J. (1980) 'A woman's mind is longer than a kitchenspoon', report on women in Sri Lanka, Colombo/Leiden: Research and Documentation Center Women and Development.

Pradhan, B. (1979) *Institutions Concerning Women in Nepal* (vol. I), Kathmandu: CEDA, Tribhuvan University.

Premi, M.K. (1985) 'Migration to cities in India', in M.S.A. Rao (ed.) *Studies in Migration – Internal and International Migration in India*, Delhi: Manohar.

—— (1990) 'India' in C.B. Nam, W.J. Serow and D.E. Sly (eds), *International Handbook on Internal Migration*, New York: Greenwood Press.

Premi, M.K. and J.A. Tom (1985) *City Characteristic, Migration, and Urban Development Policies in India*, Hawaii: East–West Population Institute.

Qureshi, M.L. (1983) *Development Planning and Women*, Islamabad: Women's Division, Government of Pakistan.

Radheyshyam, B.K., Sharma, B.K., Thakur, N. and Safui, L. (1988) 'Net weaving technology for gainful employment of rural womenfolk', paper presented at All India Workshop on Gainful Employment for Women in Fisheries Field, CIFT, Cochin (India), March.

Rahman, A. and Islam, R. (1988) 'Labour use in rural Bangladesh – an empirical

266

analysis', *The Bangladesh Development Studies*, 16 (4): 1–40.

Rahman, F. (1968) *Islam*, New York: Doubleday Books.

Rahman, M. (1975) *A Geography of Sind Province, Pakistan*, Karachi: Karachi Geographers Association.

—— (1987) 'Women and rural development in Pakistan', *Journal of Rural Studies*, 3 (3): 244–53.

Rahman, R.I. (1986) 'The wage employment market for rural women in Bangladesh', Bangladesh Institute for Development Studies, research monograph no. 6, Dhaka, Bangladesh.

Raja, A.B.T. (1988) 'Women in marine fisheries – experience of the Bay of Bengal programme', paper presented at the International Conference on Appropriate Agricultural Technologies for farm women, Indian Council of Agricultural Research in collaboration with International Rice Research Institute, Manila, Philippines, New Delhi.

Raju, S. (1980) 'School and work: variation in female literacy and employment in urban India', presented at the Annual Meeting of the Association of American Geographers, Louisville.

—— (1981) 'Sita in the city: a sociogeographical analysis of female employment in urban India', Department of Geography, Syracuse University, discussion paper no. 68.

—— (1982) 'Regional patterns of female participation in the labor force of urban India', *The Professional Geographer*, 34 (1): 42–9.

—— (1984) 'Female participation in the urban labour force: a case study of Madhya Pradesh, India', Women in Development, Michigan State University, working paper no. 59.

—— (1987) 'A socio-geographic analysis of female participation in labour force in urban India: Madhya Pradesh as an example', *Asian Profile*, 15 (3): 247–66.

—— (1988) 'Female literacy in India: the urban dimension', *Economic and Political Weekly*, 23 (44): WS 57–64.

—— (1991) 'Female labor, poverty and regional culture: the Indian experience', paper presented at the XVII Pacific Science Congress, Honolulu, Hawaii, May–June.

—— (1992) 'Women's place, women's work: the Indian experience in rural areas', paper presented at the IGU Gender and Geography Study Group's Meeting, Rutgers University, USA, August.

Raju, S. and Satish, M. (1989) 'Gender and geography: an overview from India', *Arena: Journal of Geography in Higher Education*, 13 (1): 102–4.

Rao, Bhaskara, Rao, Prakasa, B.V.L.S. and Naganna, N. 'Migration problems and policies in India', Centre for Habitat and Environmental Studies, Indian Institute of Management, Bangalore, mimeo.

Rao, M.S.A. (1978) 'Tobacco development and labour migration: planning for labour welfare and development', *Economic and Political Weekly*, 13 (29): 1183–6.

Rayappa, H.P. and Grover, D. (1978) 'Modernization and female work participation', *Demography India*, 7 (1–2): 157–74.

Reddy, C.R. (1981) 'Cotton and agricultural labourers in Berar c: 1860–1920', paper presented at Seminar on Commercialisation in Indian Agriculture at the Centre for Development Studies, Trivendrum, November 23–5.

—— (1985) 'Inter-regional variations in the incidence of male agricatural labourers', report prepared for National Bank for Agriculture and Rural Development, Centre for Development Studies, Trivendrum, India.

Reddy, D.N. (1975) 'Female work participation: a study of interstate differences, a comment', *Economic and Political Weekly*, 10 (23): 902–5.

Reddy, H.N.B. (1988) 'Development programmes for farm women in different states and their linkages with research systems', paper presented at the International Conference on Appropriate Agricultural Technologies for Farm Women, Indian Council of Agricultural Research in collaboration with International Rice Research Institute, Manila, Philippines, New Delhi.

Reddy, S.S. (1991) 'Effects of irrigation on levels of living, labour force and wages: a case study of rural Andhra Pradesh', CESS Poverty Project, Monograph no. 2, Centre for Economic and Social Studies, Hyderabad, India.

Riegelhaupt, J. (1967) 'Saloio women: an analysis of informal and formal political and economic roles of Portuguese peasant women', *Anthropological Quarterly*, 40: 109–26.

Risseeuw, C. (1987) 'Organisation and disorganisation: a case study of women coir workers in Sri Lanka', in M. Singh and A. Kelles-Viitanen (eds) *Invisible Hands*, Delhi: Sage Publications.

Rogers, S.C. (1975) 'Female forms of power and the myth of male dominance: a model of female/male interaction in peasant society', *American Ethnologist*, 2: 727–56.

—— (1978) 'Women's place: a critical review of anthropological theory', *Comparative Studies in Society and History*, 20: 123–62.

Rosenzweig, M.R. and Schultz, T.P. (1982) 'Market opportunities, genetic endowments and the intra-family distribution of resources: child survival in rural India', *American Economic Review*, 72 (4): 803–15.

Rothermund, D. (1975) *Islam in S. Asia*, Weisbaden: Franz Steiner Verlag

Rudra, A. (1989) 'Review of Women and Society in India by N. Desai and M. Krishnaraj', *Economic and Political Weekly* 24 (17): 917–18.

Ryan, J.G. and R.D. Ghodake (1984) 'Labor market behaviour in rural villages in south India: effects of season, sex and socio-economic status', in H.P. Binswanger and M.R. Rosenzweig (eds) *Contractual Arrangements, Employment and Wages in Rural Labor Markets in Asia*, Economic Growth Center Publication Series, Haven: Yale University Press.

Saeed, K. (1966) *Rural Women's Participation in Farm Operations, West Pakistan*, Lyallpur: Agricultural University Press.

Sahoo B. and Mahanty, B.K. (1978) 'Female participation in work in Orissa – an interdistrict comparison', *Indian Journal of Labour Economics*, 20 (4): 40–50.

Samarasinghe, S.W.R. de A. (1977) *Agriculture in the Peasant Sector of Sri Lanka*, Peradeniya: Ceylon Studies Seminar.

Saradamoni, K. (1987) 'Labour, and rice production: women's involvement in three states', *Economic and Political Weekly* 22 (17): WS 2–6.

—— (1988) 'Statistics on employment of women', in *Proceedings of the Second Seminar on Social Statistics*, New Delhi: Central Statistical Organisation.

Sarath, S. (1987) 'Open and hidden cash-economy of village women in Bangladesh', Bogra, Bangladesh.

—— (1988) *Economic and Political Weekly*, 23 (24), letter to editor.

Sawant, S.D. and Dewan, R. (1979) 'Rural female labour and economic development', *Economic and Political Weekly*, 24 (26): 1091–9.

Saxena, P.K. (1990) 'Employment characteristics of women: a case study of rural Andhra Pradesh', in A. Kumar (ed.) *Developing Women and Children in India*, New Delhi: Commonwealth Publishers.

Schrijvers, J. (1985) *Mothers for Life; Motherhood and Marginalization in the North Central Province of Sri Lanka*, Delft: Eburon.

Schuler, S. (1981) *The Women of Baragaon (The Status of Women in Nepal)*, 5 (II), Kathmandu: CEDA, Tribhuvan University.

Schultz, T.P. (1990) 'Women's changing participation in the labour force: a world perspective', *Economic Development and Cultural Change*, 38 (3): 457–88.

Schuth, Sister K. (1980) 'Village literacy and its correlates: a Mysore case study' in D. Sopher (ed.) *An Exploration of India: Geographical Perspective on Society and Culture*, Ithaca, New York: Cornell University Press.

Sen, A. (1983) 'Economics and the family', *Asian Development Review*, 1 (2): 14–21.

Sen, G. (1982) 'Changing definitions of women's work: a study of the Indian census', paper presented at the Golden Jubilee symposium on Women, Work and Society, Indian Statistical Institute, New Delhi.

—— (1983) 'Inter-regional aspects of the incidence of women, agricultural labourers (district-level) employment and earnings', paper presented at the Workshop on Women and Poverty, Centre for Studies in Social Sciences, Calcutta, March.

—— (1985) 'Women agricultural labourers – regional variations in incidence and employment', in D. Jain and N. Banerjee (eds.) *Tyranny of the Household*, New Delhi: Shakti Books.

—— (1987) 'Women agricultural labourers – regional variation in incidence and employment', paper presented at the National Workshop on Women in Agroculture, New Delhi, September.

Sen, G. and Sen, C. (1985) 'Women's domestic work and economic activity – results from National Sample Survey', *Economic and Political Weekly*, 20 (17): WS49–56.

Sen, I. (1988) 'Class and gender in work time allocation', *Economic and Political Weekly*, 23 (33): 1702–6.

Sevard, R.L. (1985) *Women: A World Survey*, Washington DC: World Priorities Publications.

Shah, N. (1975) 'Work participation of currently unmarried women in Pakistan: influence of socioeconomic and demographic factors', *Pakistan Development Review*, 14 (4): 469–92.

—— (1984) 'The female migrant in Pakistan', in J.T. Fawcett, S. Khoo and P. Smith (eds) *Women in the Cities of Asia: Migration and Urban Adaptation*, Boulder, Colorado: A Westview Replica Edition.

—— (ed.) (1986) *Pakistani Women: Socioeconomic and Demographic Profile*, Islamabad: Pakistan Institute of Development Economics and Hawaii: East–West Population Institute.

—— (1989) 'Female status in Pakistan: where are we now?' in K. Mahadevan (ed.) *Women and Population Dynamics*, New Delhi: Sage.

Shah, N. and Bulatao, E. (1981) 'Purdah and family planning in Pakistan', *International Family Planning Perspective*, 7 (1): 32–7.

Shah, N. and Shah, M.A. (1980) 'Trend and structure of female labour force participation in rural and urban Pakistan', in A. de Souza (ed.) *Women in Contemporary India and South Asia*, New Delhi: Indian Social Institute.

Shamim, T., Parvez, M., Abbas, S. and Parvez, S. (1985) *Battered Housewives in Pakistan*, Islamabad: Government of Pakistan.

Sharma, B.K. and Thakur, N. (1988) 'Women's participation in fisheries in Orissa – a case study', paper presented at the International Conference on Appropriate Agricultural Technologies for farm women, Indian Council of Agricultural Research in collaboration with International Rice Research Institute, Manila, Philippines, New Delhi.

Sharma, U. (1980) *Women, Work and Property in Northwest India*, London: Tavistock.

Silva, de K.M. (1977) *Sri Lanka: A Survey*, London: C. Hurst & Company.

Singh, A. (1975) 'The study of women in India: some problems in methodology', *Social Action*, 25: 341–64.

—— (1978) 'Women and the family: coping with poverty in the Bastis of Delhi', in A.

269

de Souza (ed.) *The Indian City: Poverty, Ecology and Urban Development*, Delhi: Manohar.

—— (1980) 'The study of women in South Asia: some current methodological and research issues', in A. de Souza (ed.) *Women in Contemporary India and South Asia*, Delhi: Manohar.

—— (1984) 'Rural-to-urban migration of women in India: patterns and implications' in J.T. Fawcett, S. Khoo and P. Smith (eds) *Women in the Cities of Asia: Migration and Urban Adaptation*, Boulder, Colorado: A Westview Replica Edition.

Singh, A.M. and Kelles-Viitanen, A. (1987) *Invisible Hands: Women in Home-based Production*, New Delhi, London: Sage.

Sinha, N.J. (1972) *The Indian Working Force*, Monograph 11, Vol. 1, Census of India 1961, New Delhi, India.

—— (1975) 'Female work participation: a comment', *Economic and Political Weekly*, 10 (16): 672–4.

Slocum, W.L., Akhtar, J. and Sahi, A.F. (1960) *Village Life in Lahore District*, Lahore: Social Science Research Centre, University of Panjab.

Solaiman, Md. (1988) *The Impact of New Rice Technology on Employment of Women in Bangladesh: A Case Study*, Kotbari, Comilla: Bangladesh Academy for Rural Development.

Sopher, D. (1974) 'A measure of disparity', *Professional Geographer*, 26: 389–92.

—— (1980) 'Sex disparity in Indian literacy', in D. Sopher (ed.) *An Exploration of India: Geographical Perspective on Society and Culture*, Ithaca, New York: Cornell University Press.

—— (1983) 'Female migration in monsoon Asia: notes from an Indian perspective', *Peasant Studies*, 10 (4): 289–300.

Srinivas, M.N. (1952) *Religion and Society among the Coorgs of South India*, London: Oxford University Press.

—— (1956) 'A note on sankritization or westernization', *Far Eastern Quarterly*, 15 (4): 481–96.

Standing, G. (1978) *Labour Force Participation and Development*, Geneva: ILO.

Stoler, A. (1977a) 'Class structure and female autonomy in rural Java', *Signs: Journal of Women in Culture and Society*, 3 (1): 74–89.

—— (1977b) *Class Structure and Female Autonomy in Rural Jawa, Women and National Development: Complexities of Change*, Chicago: University of Chicago Press.

Sudhakar, V.V. and Rao, V.K. (1988) 'Women's dimension in psychological research – a critique', *Economic and Political Weekly*, 23 (18): WS 20–4.

Sundaram, K. (1988) 'Inter-state variations in work force participation rates of women in India', in A.V. Jose (ed.) *Limited Options – Women Workers in Rural India*, New Delhi: ILO/ARTEP.

Sundaram, K. and Tendulkar, S. (1990) 'Toward an explanation of interregional variations in poverty and unemployment in rural India', in T.N. Srinivasan and P.K. Bardhan (eds) *Rural Poverty in South Asia*, Delhi: Oxford University Press.

Sunder, P. (1981) 'Characteristics of female employment', *Economic and Political Weekly*, 16 (19): 863–71.

Swaminathan, M. (1975) 'And miles to go', *Social Change* (1–2): 21–6.

Swaminathan, M.S. (1985) 'Imparting a rural women user's perspective to agricultural research and development', a draft paper for discussion, Workshop on Women in Rice Farming System, Los Banos.

Thorner, A. (1984) 'Women's work in colonial India, 1881–1931' in K. Ballhatchet and D. Taylor (eds) *Changing South Asia: Economy and Society*, University of London: Centre of South Asia in School of Oriental and African Studies.

Trivedi, H. (1976) *Scheduled Caste Women: Studies in Exploitation with Special Reference to Superstition, Ignorance and Poverty*, Delhi: Concept.

UNESCO (1985) *Development of Education in Asia and the Pacific: A Statistical Review*, Bangkok: UNESCO.

UNICEF (1987) 'An analysis of the situation of the children in Bangladesh', Dhaka, Bangladesh.

United Nations (1986) *Population Growth and Policies in Mega-Cities, Delhi*, New York: Department of International Economic and Social Affairs.

—— (1988) *World Demographic Estimates and Projections, 1950–2025*, New York: United Nations.

—— (1989) *World Survey on the role of Women in Development*, Vienna: UNO, Centre for Social Development and Humanitarian Affairs.

—— (1991) *The World's Women 1970-1990 Trends and Statistics*, New York: United Nations.

United Nations Development Programme (UNDP) (1975) *Regional Development in Sind: A Comprehensive Planning Report*, 1 (1), Nagoya, Japan: UNCRD.

United Nations University (1986) 'Food + Energy = Development', *Work in Progress*, 10 (1).

Unni, J. (1989) 'Changes in women's employment in rural areas, 1961–83', *Economic and Political Weekly*, 24 (17): WS 23–31.

Visaria, P. (1991) 'Discussion', ORGI/ILO–LAPTAP, national workshop on demographic and labour force analysis, New Delhi.

Visaria, P. and Minhas, B.S. (1991) 'Evolving an employment policy for the 1990s: what do the data tell us?' *Economic and Political Weekly*, 26 (15): 969–79.

Visaria, P. and Visaria, L. (1985) 'Indian households with female heads: their incidence, characteristics and level of living' in D. Jain and N. Banerjee (eds) *Tyranny of the Household*, New Delhi: Shakti Books.

Wadhera, K. (1976) *The New Breadwinners: A Study on the Situation of Young Working Women*, Delhi: Vishwa Yuvak Kendra.

Walby, S. (1990) *Theorizing Patriarchy*, Oxford: Basil Blackwell.

Wallace, B., Ahsan, R.M., Hussain, S.H. and Ahsan, E. (1987) *The Invisible Resource: Women and Work in Rural Bangladesh*, Boulder, Colorado: Westview Press.

Weiss, A. (1985) 'Women's position: socio-cultural effects of Islamicization', *Asian Survey*, 25: 863–80.

Werlhof, C.V. (1988) 'Women's work: the blind spot in the critique of political economy', in M. Mies, V. Bennholdt-Thomson and C.V. Werlhof (eds) *Women: The Last Colony*, New Delhi: Kali for Women.

Wertheim, W.F. (1954) 'Het contrapunt in de samenleving', in *Weerklank op het Werk van Jan Romein*, Liber Amicorum, Amsterdam/Antwerpen, pp. 210–17.

—— (1964) *East–West Parallels: Sociological Approaches in Modern Asia*, The Hafne: Van Hague Ltd.

Wickremeratne, L.A. (1977) 'Peasant agriculture', in de Silva, K.M. (ed.) *Sri Lanka: A Survey*, London: C.Hurst & Company.

Whittington, D., Mu, X. and Roche, R (1990) 'Calculating the value of time spent collecting water: some estimates for Ukunda, Kenya', *World Development*, 18 (2): 269–80.

Wolf, M. (1972) *Women and the Family in Rural Taiwan*, Stanford: Stanford University Press.

World Bank (1981 and 1989) *World Development Report*, Oxford: Oxford University Press.

World Bank (1989) *Women in Pakistan*, Washington DC: The World Bank.

271

—— (1990a) *Bangladesh: Strategies for Enhancing the Role of Women in Economic Development*, Washington DC: The World Bank.

—— (1990b) *Nepal: Relieving Poverty in a Resource-Scarce Economy*, Washington DC: The World Bank.

—— (1991) *Gender and Poverty in India*, Washington DC: The World Bank.

Yalman, N. (1967) *Under the Bo Tree; Studies in Caste, Kinship, and Marriage in the Interior of Ceylon*, Berkeley: University of California Press.

Zelinsky, W., Monk, J. and Hanson, S. (1982) 'Women and geography: a review and prospectus', *Progress in Human Geography*, 6 (3): 317–65.

INDEX

Acharya, M. and Bennett, L. 29, 124f, 127
Acharya, S. and Mathrani, V. 35
activity: economic definition of 2; main question on 6; non-farm 248; primary 3; principal 33; social 10
Aga Khan 30, 198, 201
Aga Khan rural support programme 204
Aga Khan University (Karachi), 207
Agarwal, Bina 99, 111, 113, 244f
age composition 1, 7, 19f, 27, 32, 40, 45, 47, 53f, 62, 70ff, 124, 182, 184, 188, 190
agricultural: assets 134; extension officers 213; implements 113; labourers 129, 192, 246; prosperity 62; surplus 113–14; workforce 18
agricultural development officer 218
agricultural development: socioeconomic consequences of 183–95
agriculturally lean season 156
agriculture: 2ff, 11, 15, 21, 27, 55, 62, 99ff, 128, 155, 197; animals in 104; capital in 102; *chena* (swidden) 31, 212ff, 217; commercial impact of 31; deviations from standard practices in 12; digging 106, 114; ecology of 12; gender in 116; in hill regions 122, 124, 196, 198; labour intensities in 13; management in 127; mechanisation of 14f, 154, 161, 206; plough 11f, 34; subsistence 62, 147, 196, 213, 217; women in 211ff, 225, 245; women's knowledge of 114, 116
alcohol 68
alcoholism 68, 116

Andhra Pradesh 12f, 20, 33, 46f, 50f, 61, 64, 71f, 250
animal: draft 154; husbandry 3, 61f, 104, 128, 143, 153, 192, 235; waste 143
animism 239
Anker, R. 8
Annamalai hills 98
Appiko 68
appropriate technology: development of 154
architecture: women students in 250
Ardener, S. 233
army service 129
artisans 160
Aryan immigration 213
Asian Regional Conference on 'Women and the Household' (1985) 137, 156
Aschenbrenner, Joyce 31, 32
ashraf: culture 24, 82; *see also* Muslim
Asia 103
Assam 46, 48, 51, 72
Asuri, T.P. and Mahadevappa, D. 32
Aswad, B. 225
autonomy 190, 194, 211, 218ff, 239, 242, 248; men's reaction to 220

Bagchi, D. 29, 240
Baghpat town 194, 195
balance of payments 248
Baltistan 204
Baltit 197
Baluchistan 65
Banerjee, N. 47
Bangladesh 1f, 4, 6, 9, 15, 18f, 33, 41, 45, 53ff, 61, 65f, 68ff, 241; martial law in 66; war 69

money-lenders 215
Monk, J. 246
Morena 140
mosque 228
mother's literacy 54
motor mechanics 115
mountain women 121ff, 196
MS study 102f, 106, 109
mud stove 146
Muslim Family Laws Ordinance
  (Pakistan, 1961) 236
Muslim, 24, 30, 36, 54f, 81f, 92, 95, 97,
  104f, 113, 135, 158, 224, 228, 241;
  converts 81; Meo 81; Shiite 229

Nagaraj, K. 34
Naidu, Sunalini 161
namaz 228, 236
nambardar 227, 229f
National Cadet Core 162
National Dairy Development Board 68
National Sample Survey (India) 3, 5, 43,
  46, 73; data of 7, 10, 13, 48, 52, 64,
  72
Nepal 1f, 4, 7, 17f, 27, 29, 41, 43, 45,
  53ff, 61, 65, 70f, 121ff
Nepal Rastra Bank 134, 136
Netherlands, The 217f
NGO 67, 69, 221
Nimkhera village 158, 161
non-wage distress work 42
non-work 4
non-workers 3
North Africa 1, 84, 148
North-Central Province (Sri Lanka)
  211, 213, 215
nutrition 69, 193

occupational distribution 55
occupations: non-primary 95; nursing
  116, 195, 235; 'suitability' of 77
opportunity cost 20
Orissa 12, 34, 46ff, 62f, 72, 101
own account production 3

Pakistan 1f, 6f, 18, 24, 33, 41, 43, 45,
  53ff, 61, 65, 70f, 135, 196ff, 224f, 232,
  234, 241, 247; Bureau of Educational
  Planning and Management of 65;
  Contraceptive Prevalence Survey of
  1; education of women in 65;
  Fertility Survey of 1, 65; Household
  Economic and Demographic surveys

in 7, 65; islamisation of 234; national
  household surveys in 8; National
  Impact Survey of 1; Public Works
  Department of 206; Rural
  Development Programme of 206;
  Women's Division of 234, 235
Palghat 102, 112
'Palitpur' village 180ff
Panikar, P.G.K. 112
Papanek, H. 138, 224, 232
Pastner, C.M. 24
patriarchy 25f, 31, 54, 156, 181, 187,
  240, 242f, 250
patronage 217
pesticides 106f, 110
petty hawkers 4
physical endurance 113
pill, contraceptive 216
pisciculture 32
planning 117, 155, 234; and women's
  welfare 117
plantation 61
planting 127, 213, 225
plough 103f, 148f, 155, 160
ploughing 104, 106, 111, 113, 148, 204;
  tribal attitude to 105; by women,
  general attitude to 105; women in
  114, 204; women's attitude to 104
pollution 135, 239
population 216
post-harvest operations 106f, 114, 127,
  153, 155f, 160, 185
potter (*kumhar*) 4, 160, 175
poultry farming 3, 32, 206f
poverty 13, 16, 21ff, 46f, 50f, 61f, 70ff,
  116, 135, 181, 193, 215ff, 236;
  absolute 122; line 9, 14, 16, 51, 72;
  ratio 72; status 135
power 116, 156; hierarchies of 235;
  relations of 213, 218, 224f, 232, 235,
  242
powerlooms 63
pregnancy 216; and physical labour 113
primary product, processing of 3
private sector 72; employment in 66
probit model 132
production sectors 55
production, means of, access to 213,
  219f
productive work 7
professional jobs 22, 66
public opinion 229
public sector 66, 72